Julian Edmund Tenison Woods

Geological Observations in South Australia

Principally in the District South-East of Adelaide

Julian Edmund Tenison Woods

Geological Observations in South Australia
Principally in the District South-East of Adelaide

ISBN/EAN: 9783743383869

Manufactured in Europe, USA, Canada, Australia, Japa

Cover: Foto ©berggeist007 / pixelio.de

Manufactured and distributed by brebook publishing software (www.brebook.com)

Julian Edmund Tenison Woods

Geological Observations in South Australia

GEOLOGICAL OBSERVATIONS IN

SOUTH AUSTRALIA:

PRINCIPALLY IN THE DISTRICT SOUTH-EAST OF ADELAIDE.

BY THE

REV. JULIAN EDMUND WOODS

F.G.S., F.R.S.V., F.P.S., &c.

Κράτος Βία τε, σφῷν μὲν ἐντολὴ Διὸς
ἔχει τέλος δὴ κοὐδὲν ἐμποδὼν ἔτι.
                        ÆSCHYLUS, *Prom. Vinct.* 13.

LONDON:
LONGMAN, GREEN, LONGMAN, ROBERTS, & GREEN.
MELBOURNE, VICTORIA: H. T. DWIGHT.
1862.

LONDON
PRINTED BY SPOTTISWOODE AND CO.
NEW-STREET SQUARE

TO

N. A. WOODS, ESQ.

AUTHOR OF

'THE PAST CAMPAIGN' 'THE PRINCE OF WALES IN CANADA' ETC.

THIS WORK

IS AFFECTIONATELY INSCRIBED BY

HIS BROTHER.

PREFACE.

THIS Work needs a few words of explanation. It has been written as much for circulation in the Colonies as for home. In the former, the number of scientific readers is comparatively few, though in no part of the world, perhaps, is a greater interest felt in matters of the kind. For this reason, the Author has entered more into detail, and given more explanations, than he would have done had the Work been intended only for men of Science. More than this — many quotations and extracts from the works of other writers on Geology are inserted. Part of them are necessary portions of the descriptions given; the rest, for the sake of comparison between what is observed in Australia and what is known in other regions. Not the least important portion of what a geological student has to acquire, is how to make use of what he reads.

In a country where so much is to be observed, it may prove useful to see how the Author has done so. This is another object of the quotations; but none have been inserted, unless as illustrating theories which might seem startling without some such support.

For the rest, the Author is sensible that there are many imperfections in the book, in palliation of which readers will kindly consider the circumstances under which it was written. All the difficulties to be surmounted need not be mentioned. Yet it may be stated, that while the missionary duties of a large district (22,000 square miles) left but little spare time, it was compiled without the assistance of any museum or library to which reference could be had, or the aid of any scientific men nearer than England whose advice would have been most useful. There may, accordingly, be many errors, and there would have been more but for the kindness of several gentlemen in connection with the Geological Society at home.

One word, in conclusion, with regard to the Engravings. The views are from photographs. The fossils, &c., are from drawings by Mr. Alexander Burkitt, of Williamstown Observatory, Melbourne (late of the Isle of Wight). This opportunity is taken of returning very grateful thanks to that gentleman for his exertions in perfecting the illustration of the Work.

PENOLA, SOUTH AUSTRALIA:
November 15, 1861.

CONTENTS.

PREFACE . . . . Page vii.

CHAPTER I.

INTRODUCTION 1

CHAPTER II.

GEOGRAPHY.

Preliminary Observations — Nature of New Country known from the Rocks—Geological Queries to be answered by Australia — General Description of Australia — Former Separation of the Continent — South Australian Ranges — The Coast of Australia — Australian Cordillera — South Australian Chain — Age of Rocks in Australia — Mineral Riches — Great Barrier Reef — General View of Australian Geology 12

CHAPTER III.

THE SOILS.

Dependence of Scenery on Geology — Description of the District — Swamps, their Localities and Peculiarities — Ridges and their Varieties — Plains — Heath and Scrub—Flora of the District — Sand and its Origin — Varieties of Soils — Honeysuckle Country — Limestone Biscuits — Broken Country — Magnesian Fermentation — Distribution of Trees — Causes favourable to their Growth — Living Inhabitants of the Swamps — Lagoons at Guichen Bay — Deposits of Bones on Banks of Swamps — In Crevices — Conclusion . . . . . Page 25

CHAPTER IV.

THE ROCKS.

Strata of the Plains — Their Uniformity — Character of the Rocks — Horizontality of the Beds — Distribution of Fossils — Sand Pipes — Native Wells — Flint Layers — Their Origin — Separation of Silica — Iron Pyrites and Rock Salt — Salt Pans — Fossil Bryozoa — Aggregation of Fossils — Age of the Beds — Corals — How deposited — Prevailing Bryozoa — Comparison of these Beds with the Remains of Coral Reefs — Difficulty as to the Nature of Coral — Extent of these Beds 58

CHAPTER V.

AN UNFINISHED CONTINENT.

Extent of the Formation — Murray Cliffs — Sturt's List of Fossils — Description of the Cliffs — Extent of the Formation in a Westerly Direction — Sturt's Account of the Formation to the North — Flinders' Description of the South — Other Observations — Boundaries to the Eastward — Tasmania — Origin of the Formation — Showing Subsidence of a large Area — Darwin's Theory — Application of this to the Mount Gambier Beds — Objections answered — Why no Remains of Atolls are found — Probably some Remains at Swede's Flat — Probable Temperature of the Sea — Geological Period — Analogies in the present State of the Earth's Crust with former Geological Epochs — Analogy of Australia to the Chalk — Retarded State of its Zoology — Bad Adaptability as a Residence for Man — Concluding Remarks 103

CHAPTER VI.

HOW THE REEF ENDED.

Cessation of Coralline Formation — Description of Upper Crag — Extent of it — Derived from an Ocean Current — Guichen

CONTENTS. xiii

Bay Beds — Absence of Fossils in them — Cape Grant Beds — Strata there described — Trap Rock and Amygdala — Similarity of Upper Beds to Upper Crag in England — Singular Formation near the Trap — Localities where the Upper Crag is found — Broken Fauna — Reefs left of Crag — Concretions not owing to Casts of Trees — Decomposition of the Rock — Blow-holes — Denudation and Upheaval — What becomes of Detritus — History of the Deposit — Denudation — Coralline Crag of Suffolk — Water-level — Deep-sea Soundings. . . . . . Page 148

CHAPTER VII.

THE REEF'S SUBSEQUENT HISTORY.

Preliminary Observations — Aspect of the Australian Coast — Sand Formation of Cornwall — Origin of Australian Sand — Its Composition — Upper Limestone and Shell Deposits — Localities in which the latter occur — Stone Hut Range — Observations on the Fauna of the Deposit — Lakes on the Coast — The Coorong — Lake Hawdon — Lake Eliza — Lake St. Clair — Lake George — Lake Bonney — German Flat — Mouth of the Murray — Upheaval of the Australian Coast — This proved from the Coast Line — From South Australian Rivers, and especially the Reedy Creek — Upheaval still going on — Periods of Rest — Six Chains of Hills — Terraces formed from old Sea-beaches — Sand Dunes not hardening into Stone — Similar Formations in Suffolk — Lake Superior and Bahia Blanca — Why generally associated with Sandstone . 181

CHAPTER VIII.

EXTINCT VOLCANOES.

Preliminary Remarks — Absence of Volcanoes from Australia — Probability of less Disturbance in Southern Hemisphere — Mount Gambier — By whom described — The Lakes — Their Peculiarities — The Valley Lake — The Punch-bowl — The Middle Lake — The Blue Lake — Mode of the various Eruptions — Volcano one of Subsidence, not Upheaval — Minerals found in the Craters — Period of the Eruption — Probability of its Extinction — Recapitulation . . 224

CHAPTER IX.

VOLCANOES — CONTINUED.

Mount Shanck — Dissimilarity of Volcanoes — Importance of describing them — Description of the Country — Well-shaped Holes — Valley — Australian Flora — Small Lake — Volcanic Bombs — The Great Cone — Remains of former Crater — How more recent Cone was formed — Its Appearance and Similarity to Vesuvius and Etna — Indentation in the Side — Evidence of former Peak — Lava Stream — Curious Mode in which it is heaped — Derived from older Crater — Cause of heaping up of Scoriæ — Parallel Instances — Connection of Mounts Gambier and Shanck — Conclusion . . . . Page 261

CHAPTER X.

THE SMALLER VOLCANOES.

Southern End of the District only volcanic — Lake Leake — Lake Edward — Craters of Subsidence — Leake's Bluff — Mount Muirhead — Mount Burr — Mount M'Intyre and Mount Edward — Line of Disturbance connected probably with Victorian Craters — Period of their Duration and the Time which has elapsed since their Extinction — Submarine Craters — Julia Percy Island — Controversy on Craters of Elevation and Subsidence — Both applicable here — Trap not always connected with Gold 282

CHAPTER XI.

CAVES.

Denudation and its Effects — Caves in general — Bones in Caves — Caves made by Fissures — How Bones came into them — Parallel Instance in South Australia — Course of Rivers in Caves — Caves in the Morea — The Katavothra — The Swede's Flat — Osseous Deposits — How Bones become preserved in Rivers — Caves which have been Dens of Animals — Kirkdale Cave — Beach Caves — Paviland Cave — Australian Caves with the Remains of Aborigines — Egress Caves — The Guacharo Caves — Other Caves — Conclusion . 299

CHAPTER XII.

CAVES.

Caves in general — Caves at Mosquito Plains — First Cave — Second Cave — Third Cave — Dried Corpse of a Native — Robertson's Parlour — Connection between it and deeper Caves — Coralline Limestone — Bones — Bones of Rodents — Other Bones — Manner in which the Caves were formed — Former Lake now drained by a Creek — Evidence of Floods — No Evidence of the Deluge — Conclusion   Page 321

CHAPTER XIII.

CAVES.

Caves — Mount Burr Caves — Vansittart's Cave — Mitchell's Cave — The Drop-Drop — Bones of a large Kangaroo — Ellis's Cave — Underground Drainage — Caves at Limestone Ridge — Other Caves — Conclusion   353

CHAPTER XIV.

Concluding Remarks   367

Appendix I.   373

Appendix II.   386

LIST OF ILLUSTRATIONS.

Caves, Mosquito Plains. Third Chamber	*frontispiece*
Map of South Australia	to face page 1
Fossils Bryozoa	,, ,, 73
Pecten	74
Retepora	ib.
Terebratula compta	ib.
Cellepora gambierensis	ib.
Spatangus Forbesii	75
Pecten coarctatus (?)	76
Cidaris	ib.
Clypeaster	77
Cast of Trochus	ib.
Echinolampus	ib.
Cast of Conus	78
,, Mitra	ib.
,, Pyrula	ib.
,, Turbo(?)	ib.
Teeth of Shark (Oxyrrhinus Woodsii)	80
Spine of Cidaris	81
Nautilus ziczac	83
Spatangus Forbesii	ib.
Cast of Turritella terebralis	ib.
Murex asper	ib.
Cellepora gambierensis	85
Branching Axis of Cellepora gambierensis	91
Fascicularia (?), South Australian Coast	187
Astræa, Ditto	ib.
Shell, Ditto	190

a

Mount Gambier, Blue Lake Crater	to face page 228
,, ,, Middle and Valley Lake Craters ,, ,,	230
Pecten coarctatus	255
Caves, Mosquito Plains. Second Chamber	to face page 325
Skull of Rodent, from Caves	336
Upper Jaw	ib.
Lower Jaw	ib.
Teeth of Upper Jaw, enlarged	ib.
,, Lower Jaw, enlarged	ib.
Kangaroo Bones	361

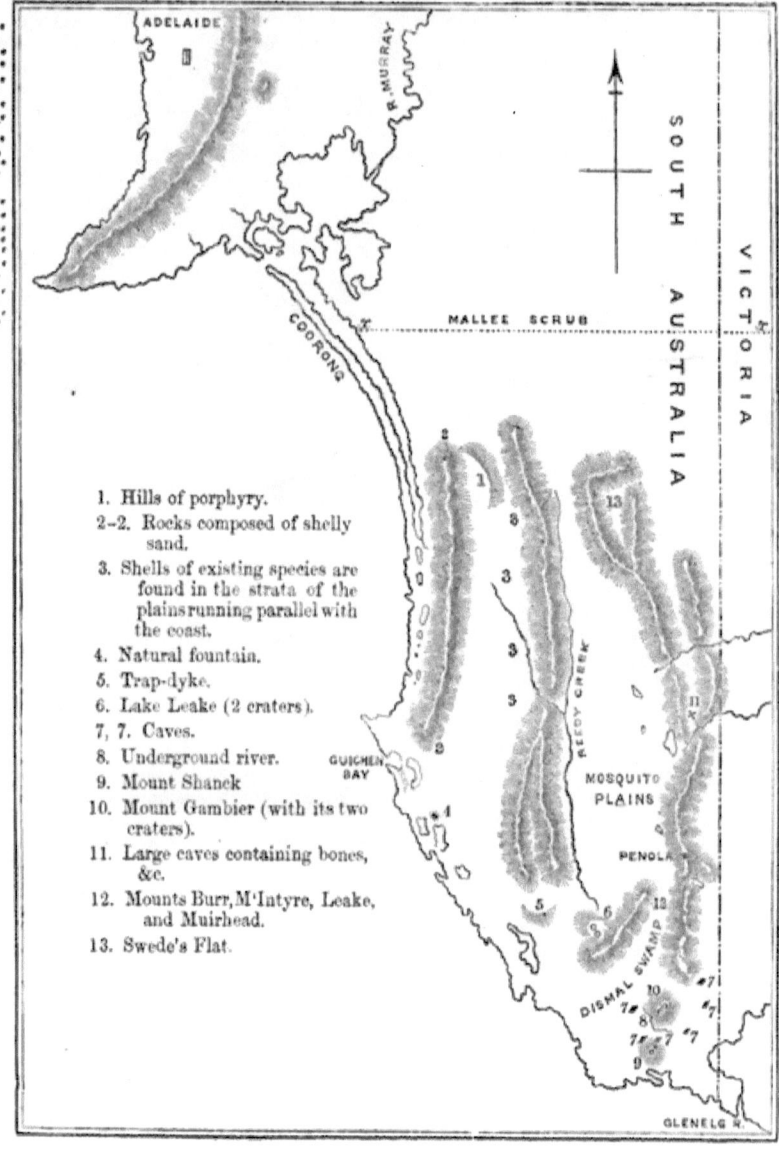

1. Hills of porphyry.
2-2. Rocks composed of shelly sand.
3. Shells of existing species are found in the strata of the plains running parallel with the coast.
4. Natural fountain.
5. Trap-dyke.
6. Lake Leake (2 craters).
7, 7. Caves.
8. Underground river.
9. Mount Shanck.
10. Mount Gambier (with its two craters).
11. Large caves containing bones, &c.
12. Mounts Burr, M'Intyre, Leake, and Muirhead.
13. Swede's Flat.

GEOLOGICAL OBSERVATIONS

IN

SOUTH AUSTRALIA

CHAPTER I.

INTRODUCTION.

MANY years ago (1683), Dr. Lister proposed to the Royal Society that a map of the soils of England should be prepared; and he urged, as a reason for it, that if it were noted how far these extended, and the limits of each soil appeared on a map, something more might be comprehended than he could possibly foresee, which would make the labour well worth the pains. 'For I am of opinion,' said he, 'such upper soil, if natural, infallibly produces such under-minerals, and, for the most part, in order; but this I leave to the industry of future times.' Geology was then in its infancy. Its only claim to the position of a science was the possession of many theories, some highly improbable, and none very consistent with the other. However, what was thought the guess of Dr. Lister was acted upon, and found to be a prophecy. A map was made, and particular soils were found to produce certain minerals, or, more correctly, certain minerals were always found associated with certain rocks, whose decomposition gave rise to particular soils. This was the first effort of geology to become practical, and already, in the distance, was seen utility. 'The industry of later times' extended these observations, and, after investigations in many places in the world which took time to accomplish, a general classification of rocks and minerals was made. Geology became thus possessed of certain principles, and, to make these of paramount utility, all that was wanted was an extensive field on which they could be applied.

A new country, whose mineral riches were unknown, was required, and this was found in Australia. Its rocks were examined, and found to correspond with similar rocks in the old country; there was an easy conclusion to be drawn, namely, that they contained similar minerals. A search was made, and was repaid by an inexhaustible supply of coal, iron[*], lead, copper, silver, and gold.

From this statement, it will appear very evident that geology is largely indebted to Australia. Not only did it give a lasting stability to principles

---

[*] The iron mines of South Australia will probably yet be found as rich as any in the world. Ores are found cropping out on the surface, within a short distance of Adelaide, from which 62 per cent. was yielded upon analysis. A ship could be loaded at the surface from where the specimen was taken. Beautiful octahedral crystals of protoxide are very common.

which were found as applicable on one side of the world as they were on the other, but perhaps more than any other country it has proved to the world that the science of geology can take the first rank as one which helps to minister to the temporal wants of man, and develop the resources of a nation.

On the other hand, however, geology has more than repaid the assistance it has received. Without going very deeply into the theory of colonization, one can easily perceive that, had Australia been only dependent on its pastoral or agricultural resources, it would have taken a long time, a very long time, to become a place of importance. Its situation is too far from Europe to have rendered its progress, under these circumstances, anything but slow and precarious. But geology has lent its hand, and given quite a different prospect. Australia promises now to be one of the most important empires that the sun will shine upon in its twenty-four hours' course round the world.

This is not a trifling thing for one science to say of itself, and its truth is very easily made evident. Look, for instance, at what Melbourne was in 1850: a poor, miserable, straggling town, with not one public building that would have done honour to a county town in England. And what have ten years done? Why, Melbourne is the wonder of the southern hemisphere. Its wharves, its Government buildings, its banks, its churches, and its parks, are evidences of prosperity at which even the fast-going Americans stop and stare with amazement. And all this is due to geology. Many will say, however, 'Don't say geology, say gold; for gold would have been eventually found without the assistance of science at all, and then this prosperity would have ensued just in the same manner.' This is true; but how long might it have been delayed? It was geology, and geology only, that led to its discovery at the particular time at which it was found. Sir Roderick Murchison, after giving some attention to the rocks of Australia, predicted, long before the discovery was made, that Australia would be found to be auriferous. The Rev. W. B. Clarke, of New South Wales, made the same observations, and it was by being urged to the matter by geologists that the Government took the matter in hand, and offered the reward which led to its discovery.

As an instance of how long the discovery would have been delayed but for science, it may be mentioned that many cases are on record of gold having been found in Victoria before it was recognised as such, and it was invariably thrown aside as either copper or iron pyrites. The author was once looking over a collection of mineral specimens collected by an old shepherd, who had a fancy for these things. Among them a piece of auriferous quartz was discovered, and, on asking the proprietor what he considered it to be, he said it was some 'copper stone' he had picked up in Victoria while shepherding, many years before.

But it is not alone in Melbourne that geology has conferred immense benefits. Look at Ballarat, at Sandhurst, and at the numerous other cities, I may call them, where, ten years ago, nothing was to be seen but a few sheep feeding. Again, in New South Wales, see what has been done for that colony by the discovery of gold fields. Not so much, perhaps, as for Victoria, but the mines are not so extensive there; it has, however, coal to make up for it, and the extensive trade of the Australian Newcastle is another proof of the temporal benefits which geology has in its power to confer. In South Australia there is ample proof of the same. About 100 miles to the north of Adelaide there is a thriving populous town, named Kooringa. This is the locality of the famous Burra Burra mine. Nothing could be more interesting than to remark the European aspect which the township, the machinery, and the population of this place present, and then to walk about two miles away, where a dreary solitary landscape, such as can only be seen in the Australian Bush, forms a singular contrast with the busy active place one has just left. What the environs are, the town itself was twenty years ago; and the change is due to geology. The copper ore was cropping out of the ground, and required no science for its discovery, but it may be doubted whether the mine would have been worth much but for working, based on geological principles, and at any rate much valuable ore would have been lost for want of a mineralogical knowledge of its value. Darwin mentions an incident in his travels which illustrates this. In some of the Chilian copper mines the copper pyrites were always thrown away, until it was pointed out by some English miners that the ore was very valuable. Now, though carbonates, not sulphurets, are the predominant ores at Kooringa, much would doubtless have been lost had not accurate scientific knowledge directed the operations.

Again, fifty miles to the south of the Burra Burra there is another thriving little township, where, much more recently, nothing of the kind was to be seen: this is the Kapunda mine, not, perhaps, equal to the one just described, but an important addition to the mineral wealth of South Australia. It has not been, as yet, certainly ascertained that the same colony does not possess coal or gold; but here again geology has shown its usefulness in directing a systematic search, and preventing useless trials where there was no chance of success.

In Tasmania, coal, and perhaps gold, give evidence of the same important position taken by one science in developing colonial resources. Many other instances might here be advanced; but what has been said is hardly consistent with the brevity intended in this introductory chapter, and, at any rate, it amply illustrates what has been advanced to establish the utility of geology. All hitherto alleged, though not immediately connected with the object of this book, will serve two important purposes. It will, in the first place, show that the science has now become so important to mankind, in bettering their social position, that all which tends to increase our knowledge in that particular branch of enquiry must be of great service. As such, it will be an apology for what follows, lest any should think its details not worth recording, or not producing sufficiently great results from the facts of which it treats. Secondly, it will give an idea of how much has been already done towards a correct knowledge of Australian geology. But on this head a little more must be said.

Of all the Australian colonies, the geology of Victoria is best known, that of New South Wales perhaps the next, and that of Tasmania next. In Victoria, the geological survey spoken of in the next chapter has been undertaken by the Government, and regular maps are in course of publication. In addition to this, the Royal Society of Victoria has among its members men of the highest scientific attainments, who are everywhere recording observations of the utmost value to the science, and, under these combined efforts, it may safely be affirmed that there are no portions of the colony whose rocks are entirely unknown.

In Tasmania, the colony has been examined by many private individuals and by the Philosophical Society of the colony. There is no Government survey, as far as I am aware; but the continued rewards offered by Government for the discovery of gold have led to an enterprising search, through which a good deal of knowledge has been obtained. Much still remains to be done.

In New South Wales, various scientific gentlemen have lent their aid to the examination of the rocks, and very little can be desired as to those in the immediate neighbourhood of Sydney. In coal districts, also, a very minute examination of the carboniferous and old red sandstone rocks has been made, and also of the rocks in the vicinity of the gold diggings. The colony, however, is so extensive that it may be many years before complete and reliable geological knowledge can be obtained.

In South Australia, nothing has been done. Though colonised as long as Port Phillip, the world, and even the residents of the colony, are quite ignorant of its geology. Certainly, the territory is very large, and a great deal of it perfectly unknown; but still the Government have never yet considered themselves justified in affording means for a scientific examination; and, unfortunately, private individuals with sufficient knowledge have never given any attention to the subject. Any searches for coal or gold that have been made have been more fruitful in negative results than in any positive information.

In 1856 a search for gold, under the direction of Mr. B. H. Babbage, resulted in nothing more than a further exploration of the colony. Many very useful and important observations were, however, made on the nature of the rocks in the districts passed through, and no doubt that gentleman's scientific accuracy in recording facts has proved of great service in giving data to carry out future operations.

In 1859, Mr. Selwyn, the Government geologist of Victoria, was invited by the South Australian Government to visit and report upon the rocks of the latter colony, with a view to its gold or coal-producing properties. In a very hurried visit, Mr. Selwyn was able to furnish little more than a mere catalogue of the rocks seen by him; but even that was of service, and certainly it was more than could have been anticipated from the short time allowed for the visit to so large a territory.

And, though so little has been done, there is no country more interesting in its formations, or more varied in its mineralogical productions, than South Australia: lofty mountains, extensive plains, sandy deserts, and inland seas, are all included in its far-stretching boundaries; with a climate like that of the south of Spain, it possesses the scenery of the Highlands in some places, while in others deserts like those of Arabia, and vying with them for bleakness, aridity, and burning heat. There are chains of salt lakes which render unprofitable a larger area than England; there are marshes and salt swamps more dank, unwholesome, and extensive, than any in the United States; there are rocky precipices, and chasms, and waterfalls to rival almost the Alps; there are extinct volcanoes of large dimensions, almost as numerous as those of Auvergne; and, finally, there are caves which exceed in magnitude the Guacharo caves of Humboldt, or in stalactites the Antiparos of the Ægean Sea.

Yet, as observed above, all these things are little known, even as existing facts, much less as illustrating scientific conclusions. To examine them all and describe them all, so as to satisfy the requirements of geology, would demand the labour, not of one alone, but the combined energies of very many learned and experienced men. This, of course, will not be obtained just now; in the meantime, any observations will be of service. With this view, these unconnected and casual observations are offered. Situated, as a missionary priest, in the large colony of Australia, there have been opportunities afforded for observation such as few could command; and thus the author has been enabled to see a very large portion of the colony, and to afford a partial observation to many facts he met with. What has been done has been recorded in this book; it will owe its chief interest, not so much from the conclusion arrived at, as from the nature of the phenomena described.

There is, however, one remark to be made before concluding this introductory chapter. A common prejudice exists now-a-days in favour of Science, which gives an unreal value to the smallest gains in its behalf. I am far from attaching an undue weight to scientific theories as such, and therefore still less to any results of my own. Though we look with the greatest pride on those things which discovery has achieved— on our telegraphs, our steam-engines, and other numerous contrivances —still their usefulness is limited and confined, and, perhaps, in reality, conferring a smaller amount of benefit than is claimed for them. If, then, we question the amount of temporal usefulness which has been awarded to geology, it has, perhaps, as much as any other science, but still of a limited and temporal kind. While these opinions are held, it will be seen that no unreasonable claims are made on behalf of what I have to relate. That they are interesting as facts I have little doubt, and because they served, in my case, as a useful employment of time which could not be otherwise occupied, they have been collected. When out in the far Bush, in the prosecution of my duties, it has been a most delightful employment, when books were unattainable, to study the great unpublished work of Nature, and it is hoped that the reader will think that the time has not been wasted.

CHAPTER II.

GEOGRAPHY.

PRELIMINARY OBSERVATIONS.—NATURE OF NEW COUNTRY KNOWN FROM THE ROCKS.—GEOLOGICAL QUERIES TO BE ANSWERED BY AUSTRALIA.—GENERAL DESCRIPTION OF AUSTRALIA.—FORMER SEPARATION OF THE CONTINENT.—SOUTH AUSTRALIAN RANGE.—THE COAST OF AUSTRALIA.—AUSTRALIAN CORDILLERA.—SOUTH AUSTRALIAN CHAIN.—AGE OF ROCKS IN AUSTRALIA, AND MINERAL RICHES.—GREAT BARRIER REEF.—GENERAL VIEW OF AUSTRALIAN GEOLOGY.

EVERY country has its history, not alone the history of what its inhabitants said and did, nor how its people lived, conspired, quarrelled, fought, and died, but a history which stretches farther back, and is buried in more remote antiquity. If it had not been so, Australia might indeed be counted the youngest as well as the least interesting of continents. She has had no people that could describe her vicissitudes, and there are no monuments left to chronicle her changes; but yet her history is written in an imperishable record. Of old, when the first explorers came upon the coast of a newly-discovered territory, the rocks, the trees, the soil, and the verdure, only spoke to them of one thing, namely, of fertility, or richness, or special adaptation to the wants of man. But now the very coast-line tells much more. Not only is the fertility or barrenness of the place itself told by the rocks, but the explorer is able to guess how far these appearances extend, and whether the country is likely to be fitted for human requirements in the present state of civilisation. Thus, for instance, if he sees granite rocks and slates on his approach, he knows that there must be mountainous ridges at no great distance — that there will be plenty of fresh water and deep soil near at hand; he knows also that mineral riches will be absent, and that every facility will be obtainable for constructing good and substantial buildings. But farther than this is the mind of the geologist carried back by the view of the granite. He pictures to himself a time when the hard stone before him was a melted fiery mass; when from the chemical laboratory of Nature new minerals were elaborated, precious stones formed, and metalliferous veins insinuated into cracks and fissures, to serve man's purposes.

So, again, if the coast be composed of chalk cliffs, geological explorers know that the interior will be gently undulating plains, but thinly timbered, that surface water will not be plentiful, that the soil will be best suited for pastoral purposes; that mineral riches will be absent; and for its history, figure to himself a white ring of breakers beating upon a circle of white sand, crowded with palm trees — a green saltwater lake in the middle of the island, contrasting strongly with the dark blue water outside — a variegated flower-show of coral animals — in fine, a marvel of fragility and strength

—of beauty and variety—a coral island or atoll, from which the chalk beds are all supposed to have been derived. In short, every stone will tell more than the mere fact of its presence. Every rock has its chronology, which can be deciphered now with ease.

And so we naturally ask, What has been the history of this vast continent of Australia, which has done so much latterly for Europe? What is the manner by which this new home came among the other countries of the world? Does it form one link in the chain of evidence found elsewhere? Does it speak of the same convulsions, changes, vicissitudes, that have attended the growth of other portions of the earth's surface? Does it, like them, speak of the dawn of creation, where simple organisation and embryonic forms told, in simple though unmistakable language, that their type and perfection—Man—was yet to come? All these questions have to be answered by the geology of Australia, and this forms the primeval history of the continent.

I am not for a moment claiming for this unpretending book the dignity of being able to answer finally all, or perhaps any, of these questions; but, nevertheless, it is meant as a contribution to the history, small in its way, but showing the continent to be no exception to the earth's previous revelations. Just as memoirs and histories of particular epochs serve to elucidate the great record of the past, so this little book will be a help towards the great results that are yet to be obtained. It is intended, therefore, in this chapter, to give a short sketch of the present knowledge of the geology of the Australian continent, because it will give the reader a notion of the relation of different portions, and the exact position of the district to be described.

I need hardly go out of my way to describe the precise locality, dimensions, and shape of the Australian continent; these are now pretty generally known. I would, however, call attention to the fact, that the outline of the continent is generally of an even unbroken outline, except in two places, the one north, and the other south; in fact, nearly opposite one another. The northern indentation is the Gulf of Carpentaria, including the straggling coast line of Arnhem's Land, with Clarence Strait, Van Diemen's Gulf, and Melville and Bathurst Islands. On the south, the indentation of the land is included in Spencer's Gulf, Gulf St. Vincent, Yorke's Peninsula, Cape Jervis, and Kangaroo Island. If we now follow the coast line from Cape Jervis in a southerly and then in an easterly one, at the River Glenelg we find there is no indentation of any size, or, at least, anything to compare with those just mentioned. This continues all around until we reach the Gulf of Carpentaria on the other side. Now, it would appear that there was formerly a separation of the continent into two halves during one of the recent tertiary periods. This separation was at or about a straight line between the Gulf of Carpentaria and the deep gulf just spoken of on the opposite or southerly side. It is hardly the place now to enter into the reasons which incline me to this opinion, more especially as it is only proposed to touch generally on the geology of the continent. There can, however, be briefly stated here a few of the facts:—At Cape Jervis a mountain range commences, which runs nearly north and south, and this is bounded on its eastern and western sides by a recent tertiary deposit. These beds will occupy a very prominent position in this volume, and so they need not be dwelt upon here any more than to say that they thin out to the eastward, very near the boundary between the two colonies, and are immediately succeeded by extinct volcanoes, bays, and altered primary rocks, which do not appear to have been covered by any tertiary sea.

To the westward of the same range the beds have been traced through the greater portion of the Great Australian Bight, until they are terminated by the primary rocks of Western Australia, which do not appear to have ever been covered by a tertiary sea. Thus we have the eastern and western sides of the continent occupied by primary rocks, and the centre by tertiary beds enclosing an abundance of recent shells. This is pretty strong presumptive evidence of their previous separation. Again, it will be mentioned, in the course of this work, that Spencer's Gulf bears most unmistakable signs of having formerly been much larger, or rather, to have been better filled by the ocean than it is at present. To the north of Spencer's Gulf there is an uninterrupted tract of waste marshy lowlands, continuing as far due north as the explorer has hitherto ventured. This has been found, wherever examined, to consist (with some small exceptions) of limestone, with recent marine shells and salt water. Many parts of this desert are sandy, while other parts are immense plains of shingles without any shells, probably portions of the ocean bed, which were too deep for the support of any animal life. Geographers are not well acquainted with the exact nature of the rocks round the Gulf of Carpentaria; but it is not unlikely that they are tertiary. The high land of Cape Yorke, on the eastern side, is known to be primary, as also the highest lands in Arnhem's Land;* and this would certainly seem to correspond with the opening for the tertiary beds at the southern gulfs. It is not, therefore, hazarding too much to say, that a sea has at no very distant period rolled between the eastern and western halves of the continent. It may be mentioned, that Yorke's Peninsula, which divides the southern gulfs, Spencer's and St. Vincent's, is composed partly of tertiary rock, and, therefore, shows its origin to have been coeval with the continent itself.

* That is to say, a ferruginous sandstone, of which the whole north coast is composed, and which is very extensively distributed over the continent. Leichardt found in it coal and plant impressions. Underneath it, occasionally, was a bed in which fossils very like Devonian types were found.

Cape Jervis is the commencement of the mountain range upon which Adelaide is built; and this is the easterly boundary of Gulf St. Vincent. Proceeding eastward from thence, or rather south-east, the mouth of the river Murray is passed; thence, unto Portland Bay, the coast is low and sandy, or containing fossiliferous and trap rock, all of them, however, belonging to the tertiary period. From Portland Bay, still keeping along towards the east, right on to Port Jackson, there is an alternation of sandy beach, tertiary trap, and Silurian rocks. The tertiary becomes more rare, once the boundary between South Australia and Victoria is passed. From Port Jackson round to Cape Yorke, in the Gulf of Carpentaria, the basis of the coast is primary.* The line of primary rocks, therefore, drawn from Cape Yorke to Port Phillip, is given as containing one vast series of rocks connected together, though, perhaps, remotely; and this, it will be seen, includes, almost in a semicircle, the whole eastern side of Australia. It is thus marked in Murchison's Map of Silurian Rocks, as distributed throughout the world. (See the last edition of his 'Siluria.')

By Sir R. Murchison it is there given as the Australian cordillera. If it is, or has been so, it has formed a remarkable exception to other cordilleras

---

* The word 'basis' is here used purposely, because there are occasional interruptions; but, from the constant reappearance of primary series, there can be no doubt that this is the general rock of the territory. The term 'primary' is not made use of in the sense in which that word was formerly accepted; those rocks are meant which are generally called palæozoic, and called generally primary or Silurian, to distinguish them from secondary, whether these be fossiliferous, metamorphic, or igenous.

throughout the world. Nearly all the cordilleras have the most gradual slopes from east to west, and their drainage, consequently, flows in an easterly direction; but in the case of Australia the slope is to the eastward, and the drainage, consequently, in an opposite course. It will be seen that the cordillera takes a westerly sweep, near the colony of Victoria, and terminates in the Australian Alps, some of which are over 6000 feet high,* and covered with snow for a large portion of the year. It is from the drainage of these mountains that the principal Australian rivers are derived, but the land through which they run is generally of a poor description, at any distance from the banks. If the continent had been formerly separated where it is here supposed it has been, the alluvial flats which are found along the beds of these rivers, and which, for so long a time, acted as an impediment to their exploration, may have been successive deltas of rivers, which are even now only very little raised above the level of the sea. In the case supposed, the Adelaide chain, which may almost be considered a second cordillera, was a chain of islands, and their further upheaval caused the rivers to take a southerly course to follow the sea,

* 'The height of these mountains is only 2400 to 4700 feet above the sea level, and even Mount Kosciusko, the loftiest of the Australian Alps, is not more than 6900, yet its position is so favourable that the view from its grassy tops sweeps over an area of 7000 square miles. The rugged and savage character of these mountains far exceeds what might be expected from their height. By far the greater part of the chain, though wooded along, is crowded by naked needles, serrated peaks, and flat crests of granite or porphyry, mingled with patches of snow.'—*Somerville's Physical Geography.*

which flowed between them and the mountains upon which Adelaide now stands.

The dimensions of the chain (which, for convenience, in this work will be called the South Australian chain) have not yet been stated. It commences properly at Cape Jervis, and continues, with varied height, until it reaches the bend of Lake Torrens, only occasionally, in its course, throwing off spurs to the right and left. This range is very detached and broken in many places, and perhaps in few parts higher than near Adelaide, where Mount Lofty rises to the height of 2100 feet above the level of the sea; but it is worthy of remark, that it is entirely disconnected with the eastern cordillera, or with the mountain systems of any of the neighbouring colonies. In examining portions of it, I have been led to think that probably they were of much greater height at one time. Indeed, it seemed to me, as far as a cursory examination could guide me, that there were very distinct marks of snow, and the action of glaciers. This would declare the range to have been once of extraordinary elevation, probably the axis of some former continent. But more minute examination must be given to the subject before anything is stated as a fact established. No fossils have been found, except at one portion of the range, about thirty miles south of Adelaide. I was informed that the fossil was a *Pentamenus Oblongus*. This would be characteristic of the lowest division of the Upper Silurian rocks. The person who found it is since deceased, so that the observation cannot be traced farther or verified, unless new discoveries are made.* With this exception, if, indeed, it can be considered such, nothing is known of the age of the rocks on this range. They are highly metamorphic, and consist principally of slates, quartzites and schists. Gold has been found in several parts of the range, although never very extensively, but the deficiency is amply made up by the immense quantities of copper, iron, lead, and silver, besides marble and various other valuable building stones. No range of hill was ever richer in beautiful varieties of minerals, and even diamonds and other precious gems have occasionally been discovered. The description of this range has been dwelt upon, because it is more immediately connected with the subject of this work. Only brief allusions will now be made to the geology of other portions of Australia.

In describing the geology of Adelaide, or South Australia, there is nothing more to say, at present, than that it is included in the geology of the range, the tertiary flats which surround it, and the district of which the description is especially undertaken in these pages. As to Victoria, its gold diggings have led to an amount of information which it would be difficult to condense, if entered upon at all.

* Since the above was written, my brother, T. A. Woods, has brought me a fossil which he found at Nuriootpa, north of Adelaide. I was long unable even to guess at its nature, it was so different from anything I had previously seen. Since then, Mr. Forbes has published his researches amid the Andes (Geological Society's Journal), and among the engravings of fossils collected from the Silurian rocks of South America, I immediately recognised the fossil of Nuriootpa. It was *Cruziana Cucurbita*, thus showing the connection of the true Antipodean beds.

Towards the close of the Introduction, it has been stated how the survey has been undertaken, and up to this time carried out. It possesses upper and lower Silurian rocks, agreeing in fossils with those of Europe, notwithstanding their wide geographical separation; it also has coal beds, which Sir R. Murchison seems to think are Oolitic, and which agree with those of Sydney by the frequent occurrence of the remarkable fern (?) *Glossopteris Browniana.**

To the geology of New South Wales I have also already alluded in the Introduction; that of Northern Australia is little known.

Western Australia is only very slightly known from recent explorations. It appears that the old red sandstone is very common, and coal is outcropping on every side; but how far this extends, or whether there are any fossils to show to what precise period the beds are to be referred, is as yet unknown. In fact, our only published source of information comes from a paper published in the Transactions of the Royal Society of Victoria by Dr. Ferd. Müller, accompanied with maps, in which paper the track of a recent exploring party in that territory is described.

A description of the Barrier Reef on the northwest side of the Australian continent will con-

---

* A great controversy is being now carried on about the age of these beds. Professor McCoy and Mr. Selwyn maintain, from the similarity of the fossils to those found by the Rev. S. Hislop at Nagùr, India, on beds known to be Oolitic, that that must be the age of the beds. The Rev. W. B. Clarke, F.G.S. of Sydney, having found the same fossils associated with true carboniferous plants, maintains that the beds are nearly the equivalents of the coal measures.

clude this chapter and the general summary of Australian geology. This is a remarkable feature, and intimately connected with some theories in the course of this work, and therefore deserves special notice. It is thus described in Darwin's work on Coral Reefs:—'The Australian Barrier Reef extends, with few interruptions, for nearly 1000 miles. Its average distance from land is between twenty and thirty miles, and in some parts from fifty to seventy. The great arm of the sea thus included is from ten to twenty-five fathoms deep, with a sandy bottom, but towards the southern end, where the reef is farther from the shore, the depth gradually increases to forty and, in some parts, to more than sixty fathoms. Capt. Flinders has described the surface of this reef as consisting of a hard white agglomerate of different kinds of coral, with rough projecting points. The outer edge is the highest part. It is traversed by narrow gullies, and at rare intervals is breached by ship channels. The sea close outside is profoundly deep, but in front of the main breaches soundings can be obtained. Some low islets have been found on the reefs.' This reef has attention called to it here, because it is closely connected with what has to be mentioned in subsequent chapters. This also is the only part of the Australian coast where there is indirect evidence of a subsidence of the land, according, at least, to the present theory of coral reefs; otherwise, generally, the whole of Australia is supposed to be slowly rising — a supposition which is more than borne out by observations to be alluded to.

In concluding this chapter, it only remains to repeat the total geological aspect of Australia. On either side, the land is composed of primary and metamorphic rocks, while the centre is an immense level tract of tertiary formations. There is no interruption to the latter, except an isolated mountain range, which forms the colony of South Australia. It has long been supposed that the central parts of the continent are below the level of the sea. Though this has not been proved, there are great probabilities of its truth. At all events, if there is any difference in favour of the land, it is exceedingly slight. These facts should be borne in mind, as having an important reference to the sequel. It has been suggested, also, that the extraordinary aridity of the inland parts of Australia is due to the coast line being so much higher than the interior, causing all winds charged with moisture from the sea to have their moisture condensed upon the coast, and thus pass dry over the interior. But the coast is hardly high enough at any place to effect this complete condensation, and practically the rains are not found so very much heavier nearer the sea. Probably the immense tract of country the clouds have to pass over before reaching the interior is a sufficient cause. But the meteorology of Australia is very peculiar, and can hardly be generalised, as yet, or compared with what occurs elsewhere, and it is certainly out of my province to pursue it further here.

CHAPTER III.

THE SOILS.

DEPENDENCE OF SCENERY ON GEOLOGY. — DESCRIPTION OF DISTRICT. — SWAMPS. — THEIR LOCALITIES AND PECULIARITIES. — RIDGES AND THEIR VARIETIES. — PLAINS. — HEATH AND SCRUB. — FLORA OF THE DISTRICT. — SAND AND ITS ORIGIN. — VARIETIES OF SOILS. — HONEYSUCKLE COUNTRY. — LIMESTONE BISCUITS. — BROKEN COUNTRY. — MAGNESIAN FERMENTATION. — DISTRIBUTION OF TREES. — CAUSES FAVOURABLE TO THEIR GROWTH. — LIVING INHABITANTS OF THE SWAMPS. — LAGOONS AT GUICHEN BAY. — DEPOSIT OF BONES ON BANKS OF SWAMPS — IN CREVICES. — CONCLUSION.

NEARLY every manual of geology commences by remarking how much the scenery of a country depends on its geological formation. The remark might be carried further. Not only does scenery depend upon geological formations, but also the appearance of the cities and towns, and even the character of the architecture. Who, for instance, does not know how much the beauty of the city of Bath depends on the excellence of the stone with which it is constructed — how a traveller in Auvergne is struck with the gloomy appearance of the town, built of scoriæ and lava? Our northern cities, again, which are built in the neighbourhood of granite quarries, have a massive style of architecture, and imperishable buildings. Look at the trim appearance of the cities of the midland counties, which have sandstone at command, or the churches of chalk flints in the south, or the neat houses of chalk in the same place. Other instances innumerable might be cited, but one more will be just mentioned, because it is a very palpable one, and because it has reference to the country now about to be described.

There are two ports on the south coast of Australia, not very far from each other; one is Portland Bay, and the other Guichen Bay. The town of the first, called Portland, is built upon a stream of basalt, which has flowed from a submarine crater at a time when the present site of Portland was beneath the waves. Nothing could be more cold and sombre than the appearance of the town, and because the houses, churches, and all the buildings are constructed with dark basalt. Robe Town, on the contrary, though situated in a most dreary bed of sand-hills, has a cheerful and picturesque appearance. It lies on a limestone tertiary formation, which supplies a pure white and durable stone for its buildings. It will be easily understood, from these instances, how much scenery will depend upon geology, if even cities do so. And more than this, the same formation generally gives rise to the same scenic effects, so that unity of rocky structure will cause unity of landscape.

In no place is this more evident than that part of South Australia called the South-eastern District. As this is the territory where most of the observations contained in this book were made, some lengthened description will be necessary. It is a territory included within the boundary between Victoria and South Australia on the one side, and the windings of the river Murray on the other. A reference to the map, at the commencement of the work, will show its precise position. The extent of this country is about 290 miles from north to south, with an average breadth of 70 miles from east to west. The whole district is remarkably level and horizontal; indeed, it may be called one extensive plain, the only exception being some ridges, which never rise more than 200 feet above it, besides extinct craters and half a dozen hills raised by trap dykes. The latter are in the southern part, and will be described hereafter. In the north, there are two or three ranges of porphyry, consisting of chains of small eminences, which run nearly east and west for about 100 miles, terminating in a volcanic district on the rivers Wimmera and Glenelg, twenty miles over the colonial boundary. Though this country is, as stated, a dead level, very little elevated above the sea, and, as far as scenery is concerned, there is the most dreary sameness, yet there is considerable unity in the nature of the plains. Thus, in the south, immediately above the craters just alluded to, nothing but immense swamps are to be seen. These occupy a fearful quantity of what might be otherwise available land. One alone, the Dismal Swamp, is of vast magnitude, stretching about thirty miles from east to west, and ten from north to south. Great as it is, it cannot be seen in one view, because it is continually run through with little island strips, or spurs of lands, which are as thickly covered with scrub and lofty trees as they can possibly be, so that the Dismal Swamp is rather a chain of marshes than one vast morass. Country of this description is not, however, confined to the southern portion of the district; it is more or less distributed over the whole. Swamps are scattered here and there, from north to south, as far as the plains last, that is, the grassy plains; and not until the 'Scrub' commences are they completely lost sight of. Some reason, however, may be assigned for their greater prevalence southward. The land rises very gradually from the sea, and, therefore, the drainage is very imperfect. There are no rivers in that part of the country, and the water can only be conveyed underground, after slow infiltration through the soft rock. The surface water will therefore collect in all the hollows, and there can be no doubt that the Dismal Swamp is a shallow depression in a very large surface of land. It is bounded on every side by very low elevations, so low, indeed, as only to be noticed by their comparative dryness at all seasons of the year.*

Whenever small swamps are isolated, they possess in this district peculiarities which deserve attention. They usually have a considerable mound

---

* In very wet seasons there is a current in the Dismal Swamp for about two months. It flows into the river Glenelg, Victoria.

or ridge of good black soil on the eastern side. This is easily accounted for. The swamps are generally densely overgrown with rushes, reeds, and a thick wiry grass. When the moisture by which they are surrounded is dried up (which occurs nearly every summer) the vegetation becomes dry and brittle, so as to be easily broken by even the wind. When in winter the rain fills these reservoirs again, westerly winds mostly prevail, and the broken mass gets drifted over to the eastern side, where it accumulates in very considerable heaps. Thus the ridges become formed: some of them are of great height, that is, considering their origin. They are, in many instances, surmounted with large gum trees, which grow excellently on this moist and fertile soil. The great size of these trees points to the antiquity of the swamps in the position in which they now are, because the ridges in which the trees are growing must have taken ages before they could give any depth of soil.*

Before going further into details, some apology may be due for what follows, as not being in keeping with the title of the work. Some facts are to be mentioned which are more botanical than geological. It is hoped, however, that their connection with the subject in hand will excuse their introduction, more especially as geology may be said to have a domain in all the sister sciences, and to be

* These ridges are frequently about 100 yards long, 50 feet high, and of considerable width, giving a very undulating character to the plains.

more or less interested in whatever may elucidate or aid them. The plan proposed in this book is to commence with the surface, and proceed through the different strata as they are found, until the lowest known here is reached. This chapter, then, though, perhaps, the least interesting, is a necessary part of what follows, and must be gone through. It could hardly be said to be a complete geological enumeration which did not show what were the peculiarities of the surface over certain descriptions of strata. Besides marshy ground, three other kinds of soil are found. They are as follow:—Ridges or low ranges of hills, with limestone cropping out; ridges of sand; and sandy plains devoid of grass. The ridges with limestone cropping out are the only elevations of any importance in the district, and these, though never of any great elevation, always border marshy country. They are always well grassed, and not thickly timbered. The most common tree on them is the *Casuarina Æquæfolia* (the Shea Oak of colonists), but others are common; the *Bursaria Spinosa* (a tree very like our privet); the *Banksia Integræfolia*, or honeysuckle; the *Eucalyptus Resinifera*, or red gum, producing a resin equal in medicinal properties to the kino gum, and the *Acacia Mollissima*, or wattle, which exudes a clear and useful gum, and is a really beautiful tree.

The soil is of a light red colour, and is evidently directly derived from the decomposition of the outcropping limestone. The flora of these ridges is poor, seldom including more than grasses, and the *Pteris Esculenta*, or common Australian fern. Nothing, however, could be more picturesque than the appearance of these limestone ridges in spring, as then their lively green colour contrasts strongly with the ordinary dead colour of Australian vegetation. In summer they are dreary enough. The grass is then dry and withered, leaving the red soil and dry rock disagreeably bare and parched. The length of some of these ridges is very remarkable. There is one which takes its origin at Penola, and continues in a northerly direction for more than fifty miles, being higher towards its northern extremity. All the way, it slopes down to the westward into an open marshy country, very sparingly timbered. There are many other ridges as long as this and parallel to it, divided from each other by plains, and this is one of the peculiar features of the district. They will be further noticed by and by, as also their caves and other curiosities.

The next kind of country to be described is that containing sandy ridges. These are generally thickly-timbered elevations, frequently forming part of those just described. They are nearly destitute of grass, but are very shrubby, rarely supporting any other tree but a stunted and irregular growth of *Eucalyptus Fabrorum*, or Stringy Bark, whose bark is invaluable for roofing, and may yet prove a superior material for ropes or coarse paper, and whose wood is the most useful that the colonist has yet found for houses, palings shingles, &c.

The kind of sand which covers these hills varies very much. In some places it is fine, mingled with loam and powdered limestone; in others, it is a mixture of rounded pink felspar and clear quartz. To describe where these sandy ridges are most common would be to indicate the whole district. They seem to occur everywhere, and run in all directions, always excepting those plains which are between the limestone ridges. Where they are of any size and continuous, their course is generally north and south, but they sometimes skirt swamps, intersect the heaths, and run round the edges of all the well-grassed country.

But the country which takes up the largest portion of the district, with the exception of the Mallee Scrub, to which it is nearly allied, is what is termed the heath. This is easily described: immense level sandy tracts, heavy and dusty in summer, and boggy in winter, supporting no grass, nor any trees but those of a stunted and worthless character, run through, here and there, with belts of short and crooked 'stringy bark,' and in all other places covered with tangled brushwood, about two feet high — these are the features of the heath. Unfortunately, as already stated, it occupies by far the larger portion of the district. In fact, if a map were coloured so as to represent only the available land, it would be seen that but a very small portion could be called such in this part of South Australia.

What is there would be seen to run in narrow strips, nearly north and south, and never more than ten miles wide. Thus, there is a strip about eight miles from the boundary, between the two colonies, which commences at the south extremity of the coast, and runs on to the Tatiara country, where it is crossed by a little patch running east and west. During nearly its whole course it is bounded on the east side by a limestone ridge, and on the west by a sandy one. Next to this strip, about sixteen miles farther west (the intervening country being filled with heath and sand hills), is another line of grassed country, not so good as the former, in consequence of its boggy nature, but of much greater extent, being about fourteen miles wide, if a ridge which runs through it be included. It runs far to the north, but the soil is very poor and light during most of its course. This is the honeysuckle country, of which more hereafter. Finally, this territory is bounded on the north and north-west by the Mallee Scrub, of which a more detailed description must be given.

This portion of the south-eastern district of South Australia, about 9000 square miles in extent, is one uninterrupted waving prairie of *Eucalyptus dumosa* (by the natives termed *Mallee*), something like a bushy willow in appearance. It commences about one hundred miles from the southern extremity of the coast, and goes on (as far as yet known), without any interruption of a different description of country, right on to the north and north-west boundary formed by the river Murray to this district. It continues on the other side of the river, but with this we have nothing to do just now. One road passes across it for about one hundred miles from the Tatiara country to Wellington Ferry, or the crossing-place of the Murray. There is also a small patch of grassy country on some porphyry ranges about twenty-six miles within its edge, but, beyond this, it is considered impenetrable. Occasionally, however, an adventurous settler has taken a few days' supply of water and provisions, and has gone fifty or sixty miles beyond the nearest settlement, but such journeys have only confirmed the idea, that the scrub is totally unfit for any purposes. There are only few places, however, where it can be even explored. The trees grow close together like reeds, and certainly not thicker, without a branch, until about fourteen feet from the ground, and so dense are they, that ten and twelve stems may be counted springing from one root, and occupy little more than a square foot of ground. Where a road has been cut through it, it appears as though there were a high wall on each side; indeed, the effect is not unlike that produced by a road through a trench.

It is strange, that while writers on Australia give so much praise to the fertility of the country, they forget to mention, that by far the larger portion of it is taken up by deserts such as this. Not only in this district, but in the whole of this South Australia, there is not a single portion of available land which is not bounded either on the north, or east, or west, by a similar desert, if the term can be applied to tracts of land producing nothing but useless stunted shrubs. The appearance of such places is very gloomy. From any eminence you see nothing but a dark brown mass of bushes, as far as the eye can reach. The soil is generally a yellow sand, and, when a patch of it is observed, it gives an air of sterility in exchange for monotony. But the outline is generally unbroken, seeming like a heaving ocean of dark waves, out of which, here and there, a tree starts up above the brushwood, making a mournful and lonely landmark. On a dull day the view is most sad, and even sunlight gives no pleasure to the view, for seldom bird or living thing ever lends a variety to the colour, while light only extends the prospect, and makes it more hopeless.

If Tartary is characterised by its steppes, America by its llanas, savannahs, and prairies, and Africa by its deserts, surely Australia has one feature peculiar to itself, and that is its 'scrubs.' Not only do they recur constantly with the same soil and the same peculiarities, but even in widely-distant districts their flora is very similar. There is something in them peculiarly Australian, which entitles them to more attention than they have received. Probably an attentive study of them will lead hereafter to more than one important result; but at least if it be asked how Australia differs from every other known continent, it may be replied, in its scrubs, and their fauna and flora.

There is a great difference, however, between what has been described as heath and the scrub; the former is generally run through with belts of stringy bark, which, in consequence of the great prevalence of bush fires in brushy country, have their trunks blackened and charred. Then there is always more tea tree, *Melaleuca paludosa*, a bushy shrub, which grows rather high; and, finally, the grass tree, *Xanthorrhœa australis*, gives a peculiar semi-tropical air to the whole. This latter tree, though frequently described in works on Australia, deserves some short notice now. It is generally a short round stem, about eighteen inches high and six in diameter, surmounted by a bundle of long, stiff, rushy grass, which droops gracefully around, and out of the centre grows something like bulrush, only longer, stronger, and much thicker in the stem. When these flower, as they do in winter, the top of the bulrush becomes covered with white stars, and a whole heath of them in flower has a very pretty appearance.

The smaller scrub, on the contrary, is a succession of clumps of the *Eucalyptus dumosa*, often bound together by a creeping plant, which makes them as impenetrable as a wall, and appearing like statues at a short distance. But there is no other plant to interrupt the growth where the mallee is thick, or, at least, of any size. In both heath and scrub, however, a shrub is common, which appears to form part of both: this is the *Banksia ornata*, a very stout shrub, with purplish flowers.

A view can seldom be obtained in the scrub except from an eminence, whereas in the heath the whole prospect lies before you in the very commencement. Both, however, have a very varied flora, though the balance is in this respect in favour of the heath. In spring, nothing can exceed the varied beauty which meets the eye on every side on both places. There is, first, the *Epacris impressa*, with its spike of campanulate white or carmine flowers; there is the *Corræa cardinalis*, something like a fuschia, tipped with yellow on the points of the corolla; there is the *Tetratheca ciliata*, a charming pink bell, and the *Dillwynia floribunda*, a tall spike, of orange papilionaceous flowers, and many others, all most abundant, and charmingly beautiful. Of course little is known of the flowers of the Mallee Scrub, but it differs from the heath in many respects. Many are found in the latter which are not seen in the former, and *vice versâ*, neither are the three varieties of Mallee found anywhere except in the scrub itself, or on its boundaries. Dr. Müller, the Government botanist, has done much towards obtaining a knowledge of the scrub flora, but of course it will not be complete until the whole has been explored.

We have now seen that one peculiarity is common to a great portion of the district described, both heath and scrub, and that is the sandy soil. It has already been observed, that the same kind of sand is not found in every place. In the Mallee it is yellow, and seems to be mixed with a great deal of clay. In some parts of the heath it is white mixed with black loam and pipe-clay, while the hills generally have the coarse-grained variety, with pink felspar. This last would appear to arise from the disintegration of the granitic or porphyritic rock; but, with the exception of a range of such rock in the Mallee Scrub, which runs from east to west, there is nothing in the neighbourhood to bear out such a supposition. It seems difficult to imagine that such immense quantities of rock could have disappeared without leaving any traces on the hills where the sand now is, or indeed any traces at all, except to the north, and a spot a little to the east of the Mosquito Plains. As, however, the pink sand is more common near the boundary, and especially so within ten or twelve miles of places where red granite now is, we must suppose it to proceed from decomposed granite, whose loose crystals have been scattered to great distances either by winds or rains. The grains are rounded and very large; probably some of the hillocks on which they are found may have been granite hills. To the sands on the heaths and sand ridges, attention must next be drawn. It varies in different places as to colour and consistency. That of the Mallee Scrub is yellowish and firm, while that of the heath is quite white and argillaceous, and like soft pipe-clay in winter. It is almost unnecessary to repeat that this sandy soil occupies more than three-fourths of the south-eastern district. It remains to suggest a few ideas as to its origin.

From what will be said in the next chapter, it will be seen that, with one or two trifling exceptions, all the rocks of the district belong to one formation: this is a tertiary limestone, containing, amid portions of coral, &c., considerable quantities of silica. The decomposition of these strata would give rise to a calcareous sand, such as that observed in the district. But not alone from this does it derive its origin. It will be shown, by and by, that, prior to the upheaval of the rocks mentioned, they were pretty generally covered with a deposit called here crag. This was thrown down from numerous ocean currents, which, with a large amount of sea-sand, carried and stratified about small fragments of shells and other marine detritus. This latter deposit has nearly entirely disappeared, and can only now be traced in a few places, where its great hardness saved it from destruction. That its removal was partly effected by aqueous agency, while the land was being slowly upheaved, there can be little doubt, but what little has remained after the process now helps to form the soil of the immense sand districts under consideration. A good instance can be seen near the coast of the extent to which the underlying rock composes the soil above. There the whole line has recently been raised from the sea; and, though the earth is of a dark colour, and contains a good deal of vegetable matter, entire shells are almost as common in it as pebbles in gravel. About two feet below the surface a hard limestone is reached full of shells of existing species.

If this theory be correct for the origin of sand, it remains to be asked, how, in some places, the soil is black (in the neighbourhood of the swamps) and in others red (on the limestone ridges), both good land, well grassed, and lightly timbered, while in other places nothing but sand is seen, though all are resting on the same kind of strata? In answering this, it must first be remembered that some allowance must be made for varieties of composition in the rocks themselves; for, though they may belong to the same age, and even be parts of the same strata, yet the local circumstances may vary. Thus, the rocks about Mount Gambier often contain iron pyrites and rock-salt minerals rarely met with in other parts of the formation. Again, in some places, the rock is of the purest white, while in others it is of a dark-red colour, with black nodules. To some such circumstance must be attributed the fact, that on limestone ridges the soil is a chocolate colour with the rock cropping out, while in the same ridge nothing but sand is discernible. But the black soil of the grassy plains requires a different explanation. In the formation of these, two things have had a material influence; namely, the lower level of the land, and the permanence of a good deal of surface water. It has been already stated that the good country (grass-supporting) runs generally in parallel bands about north and south, and it always has more elevated land upon its eastern and western sides. Now this causes a great amount of water to collect upon them, and, as there are none which are not covered with swamps and marshes, vegetable matter of every kind gets swept down by the drainage. The decomposition of this, besides the vegetation of the marshes, gives rise to the black soil. The reason why the same does not take place at the heath may be because the heath is generally on a higher level, and therefore so drained as to prevent the accumulation of vegetable manure. Indeed, this would appear to be the case, though it has not been ascertained by actual measurement. There is an instance of the kind on the river Glenelg, in Victoria. This river has cut a very steep channel for itself. The plains through which it runs are sandy heath on the higher parts; but on a large flat, which immediately borders the stream, the ground is well grassed, the soil black, and extremely heavy in winter.

It has just been said, that in all the grassy plains there are numerous swamps, and some of great extent. The lower level is shown by this, but this is not the only evidence. In winter, many square miles are in some parts completely covered with water a foot or so in depth. This is not seen in the heath. It is true that certain depressions, like the furrows of a ploughed field, cause the water to collect in small pools in winter, but there is never much of it in one place, and, even in these, the sand is darker and the ground more consistent than what is observed in the heath generally; a swamp will also be occasionally seen in the heath, but the soil close to it is black and firm.

The black soil of the plains is, from time to time,

interrupted by blocks of very white limestone. They are generally flat, rounded, of little thickness, and do not appear to be attached to the rock below. Such boulders (if the term may be applied to them for convenience) are most common on the grassy plains, and they appear to result from lime which has been washed out of the soil, and subsequently hardened on the surface by the continued evaporation of the water, in the hollows in which they are placed. Such stones are, however, quite distinct from the 'biscuits' known in the district, though their origin is owing to a similar process. These latter are found in what is called the 'honeysuckle country;' and, as that and the biscuits are very peculiar characteristics of the districts, they must be noticed separately.

The honeysuckle country, already alluded to, is, in fact, just what its name implies—extensive flats or plains growing little else but coarse rank grass and the *Banksia integrifolia*. I must be excused for turning aside for one moment to describe the tree. Though singular in appearance, it is far from being a pretty tree. A dark grey bark on a short stem, this giving rise to most irregular branches, whose smaller twigs are covered with wedge-shaped leaves, besides being studded over with flowers like a large bottle-brush, is the character of the tree. When the flowers are young they are yellow and almost pretty, but as the tree is a very long time in flower, their beauty is quite taken away by the proximity of others in various stages of decay. This makes the tree look old-looking, withered, and decidedly unsightly, more especially as age makes it more straggling, and the old seed-vessels remain on for years. The flowers produce a good deal of honey in spring, whence the colonial name. It is classed among R. Brown's *Proteaceæ*, in De Candolle's *Monochlamydeæ*. From the description given, it will be seen that a large plain, thickly studded with such trees, would be rather uninteresting. But the nature of the soil makes it far more so. It may be described as a dark grey pipe-clay, thinly grassed, covered with water in winter, and holes in summer. Cattle, by treading about it in winter and spring, leave the ground covered with their deep foot-prints, so that where the sun has baked the land the surface is as uneven as troubled waters. Riding or driving is particularly unpleasant in such a place; but what makes them even dangerous is the existence of numerous pitfalls, about a foot deep and wide. These are made by a small cray-fish, which abounds in the plains when it is covered with water. How such large holes are excavated by so small an animal, appears quite mysterious. It must be supposed that they are guided by instinct to do this, in order that they may still have a home long after the land has dried up elsewhere.

The 'biscuits,' however, are the great curiosities of this honeysuckle country. These are round flat pieces of limestone, of various sizes, as like wine biscuits as stones could possibly be. Sometimes they are small, that is, about the size of a penny-piece, covering the ground so thickly that nothing else can be seen on any side. In other places they are quite as numerous, but rather more like dumplings than biscuits, being of a large size and nearly spherical. They present everywhere a most singular appearance, more resembling a shingly beach than a plain far removed from the sea. Generally speaking, they are seldom found where there are many trees, or where the soil is dark and black. Any open space, thinly grassed, with a pipe-clay soil, seems to favour their growth the most. Persons would be led to imagine that they are the remains of coast action — in fact, that the biscuits are nothing more than what they most resemble, namely, shingles; and this would appear more probable, because, about fifteen miles west of the locality where biscuits most abound, the ground is strewn with shells of existing species, showing that it cannot be very long since the sea was rolling where they are found. But there are many reasons why this is not the true account of their origin. Shingles are seldom found except in the neighbourhood of cliffs, and there are no remains of anything of the kind here. Besides, on all the coast, even where there are cliffs, no such things as shingles are perceptible, because the limestone is of so soft and friable a nature, that if any portions become detached they are soon worn away. But the strongest reason of all against their being shingles is, that a much more satisfactory theory can be formed for their origin — one also which is grounded on a cause which is still presumed to be going on; it is as follows:—

It has been already observed that the ground is generally pitted over with little depressions, in which the remaining water collects as soon as the dry weather sets in. These are the last to dry up. In doing so, a small quantity of lime and pipe-clay (in which soil they only occur) gets hardened into a cake at the bottom. When the summer goes on, and before they are quite dry, they curl up to some extent, becoming detached from the ground, and, when quite hardened, the atmosphere and rain, during the ensuing winter, give them their rounded form. That this is the whole of the process may be easily perceived by any one who examines a few of the biscuits where they are thickly strewn, and then every stage of the process can be seen. Where there is more lime and pipe-clay, the mud (for such it is) gets detached in large fragments, and this is the cause of the big spherical masses. Where they are very small and thin, they appear to be formed from a large sheet of hardened sediment, which has cracked away and become subsequently broken small by the weather.

This explanation, simple as it may appear, would hardly be thought of at first sight. The appearance of the plains, with scattered small rounded stones, is so original and peculiar, that one might pass, wondering over them, a hundred times, without a satisfactory explanation occurring to him. And yet, how simply they tell their own story when examined; so true it is that natural phenomena are open before us like a book to read from, if we will only pay attention to every word and letter, to everything presented. However, there might arise cases where these stones would form a great puzzle to a geologist. Supposing the land were to be submerged again, and covered with another formation: after an upheaval, a geologist, finding a bed of these rounded fragments of irregular sizes, would scarcely be inclined to attribute their origin to the real cause.

It may be mentioned, that some of the 'biscuits' are covered with small mammillations or rounded protuberances, sometimes so far raised as to give them the appearance of a piece of a nullipore coral. As this is a constant form, and does not vary much in the different specimens where it is found, it must be due to some more regular cause than the weather. Unless they arose from the splashing of rain-drops from pools highly charged with lime, no other cause can be assigned; and that this is not a very unreasonable hypothesis will appear from what will be said at the end of the chapter. Some of the 'biscuits' are also rather curiously honeycombed, and appear like fragments of scoriæ, but this is clearly due to weather-wearing.

In places where there is a mixture of sand and pipe-clay, and, consequently, where the flora is of a more diversified character, possessing often many varieties of the *Eucalyptus* and *Acacia*, the soil has another remarkable peculiarity. It is known by the provincial name of 'Dead Men's Graves,' or 'Biscay Country'—names which are disagreeably expressive of the real state of things. Large tracts of this kind of soil are seen to be covered with mounds just like graves in a churchyard, only far more closely packed together than in the most thriving of our intramural cemeteries. Sometimes these mounds are two or three feet high, and then they are rounder in form; but, more commonly, they are no more than a foot in height, and then they are long and narrow, exactly within the requirements of the name they bear. They are never seen, except where there is a good deal of surface-water in winter. Those who have had an opportunity of watching them during all seasons of the year, maintain that, during the rainy season, when water has collected around them, some of the rounder portions may be seen to heave up and down in little bubbles, and the water all about has a frothy appearance, as though fermentation was going on. This I have never observed; but, as the account agrees very well with certain facts, and, moreover, corresponds with a theory I should have been inclined to propose for the origin of some of these mounds, the observation is highly probable.

The water on these flats, no matter what is the colour of the soil, is of a very milky appearance, even when in small quantities. When some of the mounds are dug into, and the rock upon which they lay is exposed to view, at a very moderate depth, it is not a limestone, but a magnesian limestone or dolomite, very compact and hard. Now the strata, as will be shown subsequently, to which these rocks belong, are full of corals, and more commonly bryozoa corallines. These possess a large quantity of magnesia, more, indeed, than any other fossil.* At any rate, I have ascertained by analysis that the quantity of magnesia, not only in the rock but in all the springs which proceed from it, is very large indeed. Now, though it is much disputed, among scientific men, what is the precise origin of dolomite, yet it is considered pretty nearly proved that it is not always produced by the same cause, and it is generally recognised that pseudomorphic action causes it, and this action takes place more frequently at the surface of rocks than anywhere else; very probably, therefore, the fermentation observed is the result of the chemical action of carbonates of magnesia and lime upon each other, and this action gives rise to those mounds whence the gas escapes and where dolomite is found. Water would appear to be the exerting cause, aided, no doubt, by the warm temperature which prevails in this climate. As to what is the nature of the action, or what kind of gas emanates from these bubbles, cannot be more than guessed,

* Forchammer analysed a great many shells to prove this, and found that, while univalves such as the *Cerithium telescopicum* and *Nautilus Pompilius* contained, respectively, only such small quantities of magnesia as 0·189, 0·118 per cent., Corals, such as the *Isis hippuris* or *Corallium nobile*, contained as much as 6·362, 2·132. Dana, also, in his investigations on dolomite, proves that coralline rocks contain sometimes more than 38 per cent. of carbonate of magnesia. Species of the *Isis* are very common in the strata here referred to.

without observation; but, as dolomite is proved to be a double salt of magnesia, lime, and carbonic acid, the proportions of which have not been exactly ascertained, probably the gas is carbonic acid, more being in the lime than is required for the salt, and becomes, consequently, liberated. But this is mere conjecture.

To prevent this theory being applied too widely, it is not meant to account for the origin of all the broken country—or even of all the mounds—which occur in the same locality; the greater proportion of them are undoubtedly due to the unequal effect of water upon the soils on which it lays. Certainly, the limestone alone cannot be considered a proper accounting cause, for these mounds are found in places where there is no limestone at all; neither do they occur everywhere, though the same rock be present, and the surface covered with water for a long time during the winter.

I am almost tempted to stray out of the limits to which this book should be confined, to speculate upon one very peculiar feature of this district, namely, the distribution of the trees. There have already been described localities which are covered with trees of only one kind, as the Mallee Scrub, the Stringy Bark, and the Honeysuckle country. Again, in the southern parts of the district, every available rise is tenanted by a great variety of trees. There are parts of the Mosquito Plains barely tenanted by any trees at all; in fact, many square miles could be picked out growing nothing but grass. What is the cause of this? As there are great doubts in my own mind whether I could throw any light upon the subject by the few observations I have made, I will not dwell further on the subject than to state one rule which appears to me to be universal here. Wherever there is an elevation, no matter how moderate, provided it be sufficient to obtain drainage, and has any kind of soil fitted to support vegetation, trees will abound, thicker, perhaps, in proportion to the shelter they receive: so true is this that the converse may almost be relied upon; and, whenever trees are found more abundantly in certain spots, these may be considered higher than the country immediately around, even though the difference may not be perceptible. Perhaps it will be a sufficient excuse for this digression to mention, that from minutiæ of the kind a good idea can only be obtained of the aspect of the district I wish to bring before my readers.

Some peculiarities of the swamps must yet be mentioned before this chapter concludes. As to their living inhabitants, their name is legion. A small fish is common to them all. Though often seen by me in the water, it has never received a closer examination. The bones of them may be found, from time to time, embedded in the mud. What is somewhat curious, these fish are found in small waterholes which have no connection with any running stream, and have been dry all the summer. The natives call the fish Lap-lap, and seem to be fond of them, though I have never seen any larger than about two inches long. There is also the cray-fish, already alluded to as occurring in the plains. The shells of these are also very frequently found embedded in the mud. There are also the usual amount of fresh-water mollusca. The species of these vary in number in different localities. Thus, for instance, in the neighbourhood of Penola, the *Limnea stagnalis*, or a species closely allied, is very common. Some of the swamps are full of them, and those which are shallow and become dry every year have the bottom whitened over with their numerous thin and transparent shells. Thus are fresh-water beds formed. On the Mosquito Plains a variety of the Paludina is most common, but their shells are never very numerous. In some of the brackish lakes above the plains the mollusca abound most, and are really a most singular feature. At a place called Lake Roy there are several salt or brackish marshes, covered with the usual tangled dank vegetation. One, however, is rather clearer than the rest, and this has all round its margin what appears at a distance not unlike a bank of sand. It is entirely composed of small shells of fresh-water mollusca, being, as I believe, a species of *Paludina*. None of them are more than about half an inch in length, and very few attain even that length, so the numbers may be guessed when they form a deposit several feet thick many yards round a lake certainly more than half a mile broad and long. There is a singular scarcity of land shells all over this district. I never met with any but a few of one species of *Succinea* in the sand-drift near Guichen Bay. There are a great many varieties of the minute crustacean Cypris in every lake, but more especially in the water which has collected at the bottom of caves. In some of these little subterranean lakes at Mount Gambier the water actually looks turbid from their immense numbers, and, if a little of the mud is taken from the bottom and examined, it will be found to consist almost entirely of their shells. The *Cyclops vulgaris* is also extremely common here as elsewhere, and serves in like manner to give the water a troubled appearance, so numerous does it at length become in some of the swamps. Many other of their living inhabitants might here be enumerated, but they would, from the above specimen, form but little exception to what are found in fresh and salt water. As yet, my attention has very little diverted to the microscopic forms occurring in them, though I have no doubt there is a very rich field open to the investigator, not only for *Infusoriæ*, properly so called, but for many new varieties of *Desimadæ* and *Diatomaceæ*. The sands also, which are so very plentiful in this district, have shown that in many instances they are entirely composed of the siliceous shields of *Diatomaceæ*. In one cave, indeed, where the sand had the consistence of white flour, and appeared somewhat different from the arenaceous deposits in other localities, it was found to be wholly composed of the frustules of these minute beings.

It is worth while here to mention two other swamps which are remarkable for their deposits, especially as there will be no other place in this work where they can be conveniently noticed. One is a fresh-water lagoon, not very far from Guichen Bay. It is shallow and dries annually, and then its bottom and sides seem to be encrusted with a white efflorescence very much like salt. Many persons who have been constantly passing this lake assured me that the lake water was salt, and that the white deposit was the crystallised salt. On examination, however, the water was found to be fresh, and the deposit no other than an amazingly thick growth, of a very white variety of the common *Chara*.* The quantity of this little plant is enormous, since it covered the whole bottom of the lagoon, and the banks are formed of broken portions, some inches in thickness. The other lake worthy of notice is close to the sea, very deep, and with its banks crowded with a rank vegetation. One of the shrubs found there occurs in no other part of this colony, so far to the east, excepting on a few spots near the shore: this is the *Corethrostylis Schultzenii*. The waters of the lake are very salt and bitter, and of a dark yellow colour. It is so highly charged with lime and magnesia that pieces of wood and roots of trees plunged into it

---

* Lake Wallace, another fine sheet of water, is covered with a very thick growth of *Vallisneria spiralis*.

become in a very short time enamelled with limestone. There can be little doubt that, if left long enough, it would change the texture of the internal parts of the wood and partly petrify it. The bottom is of very soft clay, so deep and finely levigated that a pole may be plunged ten or twelve feet into it with a very small pressure. What is very singular, the lake, though scarcely a hundred yards across, abounds with fish. I have been informed that it was the custom of persons fishing on the coast to place very small ones in this lake. Some of them are now very large, and how they exist in water so highly impregnated with salt is certainly curious.

In mentioning that some of the swamps have high banks of vegetable mould round them, it should be stated that these have sometimes been found by me to contain bones of small marsupiala of existing species, at a very small depth below the surface. It is easy to understand how these become embedded there. During the summer, the swamp being dried up, the bottom becomes a place of resort for vegetable-feeding marsupials, as the grass is longer and greener there than elsewhere. Here the smaller ones frequently fall a prey to the eagles (*Aquila fucosa*) and other birds of prey, unfortunately too common in the district, who leave the bones after devouring their victims. I have frequently met with remains of this kind when riding across the dry swamps. The westerly winds in winter heap these up with the other detritus on to the mounds. Bones of larger animals have also been found in the same sort of place, but not in this district. At Lake Colac, not far from Geelong, on a huge mound, near the lake, a great quantity of very large bones were found, as I am informed by a gentleman who resided near there. Probably they were bones of that large kangaroo called the Euro, which is only now found in the very far north, but which, from bones in my possession, obtained from caves in this district, must, at one time, have flourished as far south as this latitude, which is little, if any, to the north of Lake Colac. They might, however, have been bones of the extinct marsupials, similar to those found on the Hunter River, New South Wales.

The animals to which the bones belonged very likely perished in the soft mud of the lake, in attempting to get water. Similar instances are of daily occurrence. Persons unacquainted with the locality are often astonished at the quantity of bones of cattle, sheep, and horses which are round some of the deep swamps, which must have accumulated during the last seventeen years, as the district has not been settled upon longer. But in every dry season the mystery is solved. The poor animals are driven by thirst to go far into the mud before they can reach the water, and, being unable to extricate themselves from thence, perish, and leave their bones to be embedded or washed up during the ensuing winter. Kangaroo, wombats, &c., will only be found near watering-places in summer, and the latter, though they burrow in dry places, will often go a very long distance in search of water. It is curious to remark, that though I have seen hundreds of kangaroo in different places, at all seasons of the year, I have never but in one instance seen any of them drinking.

This chapter has been devoted to the description of those formations where geological relics may be found from what is now taking place. There is no other deposit of bones found, except in the crevices of the limestone rock, where, in consequence of the drifting of the winds, or by the force of rains, little masses of relics of existing animals may be looked for. One instance will suffice:—At a round water-hole, near Mount Gambier, caused by the falling in of the rock, little drifts of what look like roots may be found on the ledges of the strata above the water; they are composed of broken fossils from the rock above, withered leaves, roots, and the bones of native cats (*Dasyurus Maugii*), which live in these crevices and prey on birds, &c., and some birds, in which those of the young of the native magpie (*Gymnorrhina leuconota*) are the most common. In some cases, the water charged with lime drips from above, and forms these into a conglomerate, which would be easily mistaken for part of the limestone strata, if the observation were not carefully made.

This rather lengthy chapter must now be brought to a close, and we pass on to the deposit of coralline rock, to which so many references have been made.

NOTE TO CHAPTER III.

There is a curious circumstance connected with these swamps which have an underground drainage, which, in any other than a new country, would surely have been invested with some ghostly legend. Every evening, during spring and the early part of summer, distant groanings are heard, like the lowing of a large herd of cattle, and very resonant, near a few swamps, such, for instance, as that situated near Mr. Donald M'Arthur's station, Limestone Ridge. Generally, three such echoing sounds are heard, and then about half an hour's repose. I believe the sounds are entirely due to a column of air resisting a column of water, which is draining through the limestone, and finally, being driven back or forwards, according to the periodical increase of the weight of water. To one ignorant of the cause, the sounds are mournful and startling in the extreme, and they are not heard in the day, probably because there are so many other sounds of cattle, &c., to mingle and be confused with them.

On the coast also, where there are sand-stones (to be subsequently spoken of), noises like distant artillery are heard on windy days. Dr. Phipson mentions these sounds as being very common on the sandy parts of the coast of England, and is at a loss to assign a cause. It seems, however, to be in some way connected with large collections of sand. Sturt mentions that when in the Australian desert, surrounded by the high hills of red sand of that inhospitable country, he was startled one morning by hearing a loud, clear, reverberating explosion, like the booming of artillery. The next morning he heard it again. The mornings were calm and clear, and they were at least 600 miles from the settled districts. My brother (Mr. T. A. Woods), when at Mount Serle, in the horseshoe of Lake Torrens, which is a very sandy desert, has frequently heard the same loud boomings on fine clear days. They seemed to come with a startling echo from the sandhills, and reverberated for a long time among the hills. Mitchell and Sturt have observed the same thing in other parts of Australia. May the cause not be similar to that which makes the sand musical at Eigg (see Hugh Miller's 'Cruise of the *Betsy*,' chap. iv.), the sonorous moving sand at Reg Rawan, Cabul, and the thundering sand of Jabel Nablous, in Arabia Petræa? In the latter case, the mere falling of the sand on the rock beneath made a sound like distant thunder, and caused the rocks to vibrate. The ultimate cause is quite unexplained.

CHAPTER IV.

THE ROCKS.

STRATA OF THE PLAIN. — THEIR UNIFORMITY. — CHARACTER OF THE ROCKS. — HORIZONTALITY OF THE BEDS. — DISTRIBUTION OF FOSSILS. — SAND PIPES. — NATIVE WELLS. — FLINT LAYERS. — THEIR ORIGIN. — SEPARATION OF SILICA. — IRON PYRITES AND ROCK SALT. — SALT PANS. — FOSSIL BRYOZOA. — AGGREGATION OF FOSSILS. — AGE OF THE BEDS. — CORALS. — HOW DEPOSITED. — PREVAILING BRYOZOA. — COMPARISON OF THESE BEDS WITH REMAINS OF CORAL REEFS. — DIFFICULTY AS TO THE NATURE OF THE CORAL. — EXTENT OF THESE BEDS.

HAVING occupied some considerable space in the description of what is seen on the surface of this part of South Australia, that is to say, the physical features, the soils and their products, it now remains to describe what peculiarities are next in succession, thus bringing us to the consideration of the rocks. The district has already been described as an immense plain, with very few elevations of any kind, and certainly none that can be considered hills. Under such circumstances, very great uniformity in the underlying strata must be expected, and, therefore, very little geological variety; for abrupt transitions from the rocks of one period to those of others far removed in age are only found in hilly countries, where there has been much upheaval denudation and general disturbance. Near Adelaide, for instance, where bald and rugged hills are common, the most sudden changes are observed in the nature of the strata. Thus, Mount Lofty is a metamorphic rock, continually changing in its gullies to slates, porphyries, schists, and black limestone, and these, again, are often covered with tertiary limestone. It must be stated, however, that uniformity is the rule rather than the exception in Australia, and perhaps there may have been less disturbance altogether in the southern than in the northern hemisphere. At all events, the locality now under consideration is of a very uniform character, there being a large territory occupied by one formation, and this without alteration of level, break, or interruption.

Of the large area spoken of in the previous chapter, and covering many thousand square miles, the series of strata all belong to one period and have been formed under the same circumstances, and, in all probability, during the same geological period. There are only one or two exceptions to this continuity, and these are not breaks but patches, where a more modern deposit lies above the older and larger strata. The lowest and oldest will be first described, because they are the most important, and a knowledge of their characters enables us much better to understand the circumstances under which the others are formed. The nature of the rock now first deserves attention.

At about four feet below the surface (sometimes

less, though seldom more,) a brittle white limestone is met with. It is generally friable, and, being much decomposed, contains no fossils. By decomposition is meant that it is as fine as flour when dry, and run through in every direction with little veins of clay, which are very like in colour and consistence to the chocolate soil above the rock. This goes occasionally to some thirty feet below the surface, though sometimes not so far. It is, at times, entirely absent, then regular signs of stratification occur with the commencement of organic remains. Here the rock changes its nature; instead of being loose and friable, it is hard and close. It is quite white, resembling chalk, being easily cut with a saw, and, though rather soft, answers excellently for building purposes, giving rise (from the easy manner in which it is worked) to a more decorated style of architecture than is usually met with in the bush. There are no marks of stratification in small portions of the stone, but where a large section of the beds is seen such traces are very distinct. It is there observed that there are strata occurring every fourteen feet with great regularity, and in nearly every case parallel with the horizon.

This latter fact shows that there has been no violent upheaval. The structure of the strata is not always the same in every case. In general, where the fossils are large, and containing many bivalve shells, the stone is very hard and durable; but where there are only bryozon fossils, or foraminiferous remains, the stone is mere powder when disturbed. In the latter case, the only thing which gives it the least consistency is the occurrence of twisted concretions, whose appearance and origin are best treated of in a subsequent chapter.

It is generally remarked that the different strata preserve distinct characters. Either the bivalve shells will predominate, and then the whole stratum be firm and hard, or else they will be entirely wanting, and the stratum soft and powdery. Sometimes a stratum will be found with a character more or less between the two, but this is unusual. Without stopping now to examine the nature of the organic remains, any more than to state that the most passing examination shows the rock to be nothing else than a mass of fossils cemented together, and even the dust is seen by the microscope to teem with the relics of life, several other peculiarities of the strata have now to be mentioned.

The stone has been described as very like chalk. This resemblance might be ascribed to the whole formation. Many analogies might be here mentioned, but two things which make the likeness very striking are here selected: these are 'sand-pipes' and layers of black and white chalk flints. The first are well known to those who have examined any of the chalk-beds of Europe. They are described by Sir C. Lyell as 'deep hollows of a cylindrical form, found penetrating the white chalk, and filled with sand and gravel.' They all taper downwards, and end in a point. As a general rule, sand and pebbles occupy the central parts of each pipe, while the sides and bottom are lined with clay. The strata under consideration are full of these. When a section of the rocks, they are observed in great numbers, some about eighteen inches in width, and not extending below the first stratum; others are wider than two feet, and going to a considerable depth. Sir Charles Lyell supposes these to have arisen, in the first instance, by coast action, and to have been subsequently defined by the action of water charged with carbonic acid, which would dissolve and decompose the limestone. The clay in the stone would be derived partly from the disintegration of the stone and partly from above. This is the theory received at present. Two circumstances observed here tend to bear out its truth. Whenever the limestone is decomposed, it gives rise to clay very much like what is found in these pipes, and their cylindrical form is due to a process which is seen at present in operation on the coast. In the latter place, wherever the sea has much action on the rocks, the pipes are seen in every stage of their formation; sometimes as a mere basin, where the water collects and dissolves the limestone; in places as little wells of varying depths, and in this and every other instance lined with a coating of laminated limestone all round, so that the surface of the outermost coat forms a continuous lining to the pipe. It does not appear why these should form so rapidly as they evidently do. In other respects, the phenomena may be explained by supposing the sea-water to dissolve the limestone and re-precipitate it with some of its own salts round its sides during the evaporation of the water. The solvent power of the water, and not the violence with which it is dashed into these pipes, can be alone looked to for an explanation, for the process goes on in those which are only filled by rain. There will be occasion, by and by, to mention these pipes again, in connection with the so-called fossil-trees so common on the Australian coast.

Next to these pipes in resemblance and in interest are what are termed the 'native wells.' These are round hollow tubes, going to a great depth (generally the water-level), three or four feet wide, and bearing considerable likeness to an artificial well. Between Mount Gambier and Mount Shanck the ground is studded all over with these wells; some are as wide as five feet, and have been sounded with a hundred feet of line without finding bottom, though water has been obtained at depths varying from sixty to ninety feet. Their origin must be different from the sand pipes. They are, perhaps, connected with caves or reservoirs of water underneath, and when, in consequence of some original depression in the ground, the water was able to rest upon the limestone and decompose it, a passage was easily formed to the water-level below. This is merely offered as a suggestion.

One fact (probably of not much importance either way) may be mentioned:— The 'native wells,' as they are called, are only seen where caves are common, and where the ground in the vicinity sounds hollow on percussion. At a cave in the township of Mount Gambier, where a long subterranean passage is filled with water, several of these natural wells lead down to it, and one or two may be noticed where a section of the strata is seen, and the decomposing rock in one well has not yet reached the cave. Though this is broad, it is filled with clay, while another narrower has bored through and is empty.

The next point of resemblance between these strata and the chalk is the occurrence of layers of flint similar to those met with in the latter formation. Their characteristics are almost the same as those found in Europe, being most frequently black, containing sponges, corals, spiniferites, &c., imbedded and occurring in layers. Some of them are large and rounded; but at one place on the coast (Port M'Donnell) they occur in sheets of very great extent, and about two or three inches thick, and are quarried and used as flags. It is stated by those who have had much experience as well-sinkers in this neighbourhood, that a layer of them is always found immediately above the water level, and, as far as my own observation goes, the strata are generally parallel with those layers.

It is generally admitted, by those who have written on the subject of the chalk flints, that they are derived from the filtration of silica through the strata, but it has not been satisfactorily explained why they should occur in layers, nor what is the nature of the process by which silica acid becomes free. Perhaps one circumstance has been overlooked in the theory of filtering through. Though water will dissolve a certain amount of silica, it will much more readily dissolve carbonates of lime and magnesia. Supposing each to have been dissolved in small proportions, silica should have been the last to filter through, because its specific gravity is less than either of the other salts. It may be that chemical action goes on between the carbonates of lime and magnesia, in the formation of dolomite, which contains a very small portion of silica, and that, mechanically mixed,* this leaves the silica to aggregate by itself. Sea water does not contain much of the latter mineral, and coral scarcely more. Sponges, however, and their spicula, contain a great deal, and the shells of infusoria are nearly entirely composed of it. Both are most abundant in the sands on the coast, and may probably have been so where the flints are found; in fact, the latter often contain fossil sponges as a nucleus. That chemical changes do take place in the rocks, and lead to an association of minerals of one kind, is seen, from another part of the same formation, at Portland, Victoria, where there are many veins of soapstone occurring in the strata, and running through them without

* Forchammer.

any appearance of a break or dislocation in the rock.

It may be remarked, as serving to elucidate the origin of such nodules, that they are only found where the beds are proved to have been derived from coralline and coral beds, rich in fossil corals, bryozoa, sponges, and infusoriæ. The beds now described are tertiary, certainly not the only ones which contain flint — features at one time supposed to be confined to the chalk. There can be no doubt that observation is much wanted as to the manner in which silica may be deposited by filtration. Quartz veins of segregation are familiar to every one, and there will be yet occasion to describe instances of segregation which will prove that a great deal has yet to be learned before a comprehensive theory can be formed, which account for the many and various facts met with. Mr. Sorby's important paper on the microscopic structure of crystals (read before the Geological Society, December 2, 1857) is something done in this particular. As this has shown where the field lies, its value can scarcely be overrated. It has thrown light where all before was gloomy conjecture. Though any theory would as yet be premature, yet geologists can already see which way facts tend, and at any rate there is something to work upon — an opening for the thin edge of the inductive wedge.

The fact of the strata being similar in composition to the chalk, besides containing flints and sand pipes, might mislead superficial observers, who had not examined the fossil, to think the two deposits identical. This is one of the many examples of the fallacy of laying any stress on such characteristics. Physical properties (if the term may be allowed) may produce the same results in strata vastly remote from each other. In all cases where such grounds are adopted as part of a classification it would be well to be sure that peculiar circumstances may not produce such resemblances where dissimilarity exists in every other respect. Had this precaution been adopted by the earlier geologists, the science would not now be encumbered with such ambiguous terms as red sandstone, oolite, mountain limestone, &c., &c. It may even be doubted whether, at the present day, definitions of deposits included in any geological period are not encumbered too much with particulars of structure. This is a great embarrassment to the young student in remote countries. Fossils should alone be relied upon. This is one of the many instances of the superiority of a natural system of classification, even although an artificial one may be useful at times and always easy.

In addition to flint layers, other minerals are found: they are iron pyrites and rock salt. The first is uncommon; it is only found near flints, which are much reddened by the oxide of iron. It hangs down, like small stalactites, in cavities in the rock. The colour outside is a dark-reddish brown, the fracture showing yellow crystals of feeble brilliancy. The rock salt is hard, massive, and opaque, so purely white as to be easily mistaken for quartz. Very little has been found. It occurs on the upper side of the great divisions into which the strata are divided. Being found in cakes, its external aspect is rather singular, such, indeed, as to make it with difficulty distinguishable from the rock by which it is surrounded. The upper side is covered over with a crust of dust, like the powdery rock, which looks as if it had fallen on the salt at a time when it was in a soft state, and to have become subsequently agglutinated by drying on. The appearance is very like the surface of dried glue on which sawdust had been sprinkled while warm. No doubt the cause has been the deliquescent nature, which melts in damp moist weather, and hardens again in drier seasons. Its existence here is not easily accounted for. Whatever may be the theories with regard to rock salt when it occurs in large beds or pans, or when it is associated with volcanic emanations, as in the Carpathian Mountains, certainly no such theories will apply here. There is nothing in the rock where it occurs different from places where no rock salt is found — no traces of anything like an immense evaporating surface, and so its presence is simply an enigma. Perhaps sea water might have flowed down the cracks of the rock, and have become evaporated as fast as it was supplied; but it must be owned that nothing has been seen to bear out the idea. In this case, it would be too gratuitous an hypothesis to adopt the theory of chemical action as a cause, acting in a similar manner to that in which dolomites, pyrites, and flints are produced.

SALT PANS, OR 'SALINAS.'

It may be mentioned here, though more properly belonging to another chapter, that salt pans, or 'salinas,' are not uncommon in the district. These are immense basins or swamps, filled with brine in the rainy season, and in summer the water evaporates, leaving a thick crust of salt in the bottom, white and glistening, giving the appearance of the ground being covered with snow. Such places as these serve to supply the country round with salt. It is not of a good description, being generally very coarse and dirty, somewhat bitter in taste, and always containing a small admixture of sulphates of lime and magnesia. Round most of the lakes there is a border of whitish mud, of a very fetid odour, and in this large crystals of gypsum and natron occur.

There is no difficulty in accounting for these deposits. The land, indeed the whole coast, is known to be slowly rising from the sea, and these pans have been, in turn, depressions near the coast, subject to occasional inundations from the ocean. Of course the continued evaporation of such casual additions would cause a great deposit of salt in the bottom of the lake, which would remain long after the time where upheaval had placed them beyond the reach of the sea. Even a large body of salt can be accounted for without supposing any communication with the sea after a slight upheaval. There is a large lake, known as Lake Eliza, not very far from Guichen Bay. It has no communication with the ocean, and has once been much deeper than it is at present. It has become evaporated into an immense shallow pan. The water is excessively salt and buoyant. No one can doubt that by the time the whole has evaporated there will be an immense quantity of salt, arising from the large body of sea water which has dried up.

After having given this much attention to the substances occurring in the rock under consideration, let us return to the rock itself. It is, as before observed, of various consistency, sometimes fine grained, and containing no fossils; at other times, exceedingly rich in them. Where there are none, the stone seems to be nearly a hardened lime paste, such as might have been derived from the comminution of small Corallines and Foraminifera. Doubtless an Ehrenberg might discover vestiges of an animal life in them as distinct from the microscopic world of Europe as they themselves are from the large fossils by which they are surrounded. Very delicate appliances for microscopic investigation were not within my reach, but what has come under notice shall be here specified. It must first be mentioned, some fossils of this deposit were sent home to the Geological Society in 1859. From the dust accompanying some of them, T. Rupert Jones, Esq., F.G.S., was enabled to procure many *Foraminifera*. He stated that they were mostly of Pleiocene origin, and showed a sea bottom, which must have been covered with at least between 200 and 300 fathoms of water. This, it may be presumed, is only on the supposition that the animals lived where their remains were found, and were not brought from a distance. It will be easy to show hereafter that this supposition cannot be applied here.* For the purpose of searching *Foraminifera*, the finest dust was taken and sifted through muslin. The dust which came through, on being placed under the microscope, appeared full of microscopic fossils. It would be useless for me to attempt anything like a classification without any museum to which I could have recourse for the comparison of specimens; it will be sufficient, however, to mention, that, though the dust was composed entirely of fossils, they all seemed to belong to D'Orbigny's order of *Monostega*, that is, of shells comprised in a single segment or chamber. If there was any exception to this rule, it was in the occurrence of what appeared to me to be a small Cristellaria; there may have been other fossils, but my skill in microscopy was not sufficient to enable me to detect them. With the larger dust which remained after the sifting the variety was much greater. In addition to inconceivably small fragments of *Bryozoa*, almost every variety of *Foraminifera* might be found; in fact, the stone was a mass of them. They were

* Since writing the above, the following remarks have been made in the 'Quarterly Journal of the Geological Society,' on the subject of some fossils sent home by me, by Professor T. Rupert Jones, F.G.S. :—
'A small portion of this deposit has yielded several *Foraminifera*, namely, *Polymorphina lactea*, *Textularia pygmæa*, *T. agglutinans*, *Globigerina bulloïdes*, *Cassidulina oblonga*, *Rosalina Bertholetiana*, *Rotalalia Ungeriana*, *R. Haidingerii*, *R. reticulata*, *R.* (*Anomalina*) *rotula* (rare — the rest were marked more or less common). The above-named Rhizopods exist at the present day, and for the most part, in rather deep water, at from 200 to 300 fathoms. It would hence appear that the fragmentary *Bryozoa* forming the mass of the deposit were washed down from a higher zone, and mingled with the *Foraminifera* inhabiting deep water.'—*Geological Society's Journal*, November 1859.

opaque, but it seemed to me that the foramina were visible on the surface in the shape of minute granulations.

The coarser the dust the more numerous the fossils, and the greater the variety. It is impossible to say whether the species were similar to those found in the chalk and other strata elsewhere. It is very probable, that though so great a diversity exists between the fauna of this and other countries, an equal difference does not exist in the microscopic animalcules.* The question, however, is one well worth solution. If this book should fall into the hands of any who are willing to pursue this subject, and have the opportunity, they have but to take a piece of Mount Gambier stone and brush it with water. The resulting dust, when dried and sifted, will afford ample material for the examination, and there cannot be a more entertaining employment and amusement than to inspect the innumerable varieties of form which a small fragment of stone contains within itself — to trace out the perfect remains of these tiny organisations, and to reflect that Time, which has crumbled the bones of mighty kings, and torn down cities, has spared these atoms, the date of whose existence is so

---

* The Foraminifera are the oldest fossils in geology. Some are found *specifically identical* with those occurring in Palæozoic deposits, and the species found in the Arctic regions, in the Gulf Stream, and in Australia, do not differ from each other. The large *Operculina arabica* and *Globigerina bulloïdes* are the most common at Mount Gambier. They are so large as to be hardly microscopic. See engravings.

FOSSIL BRYOZOA.

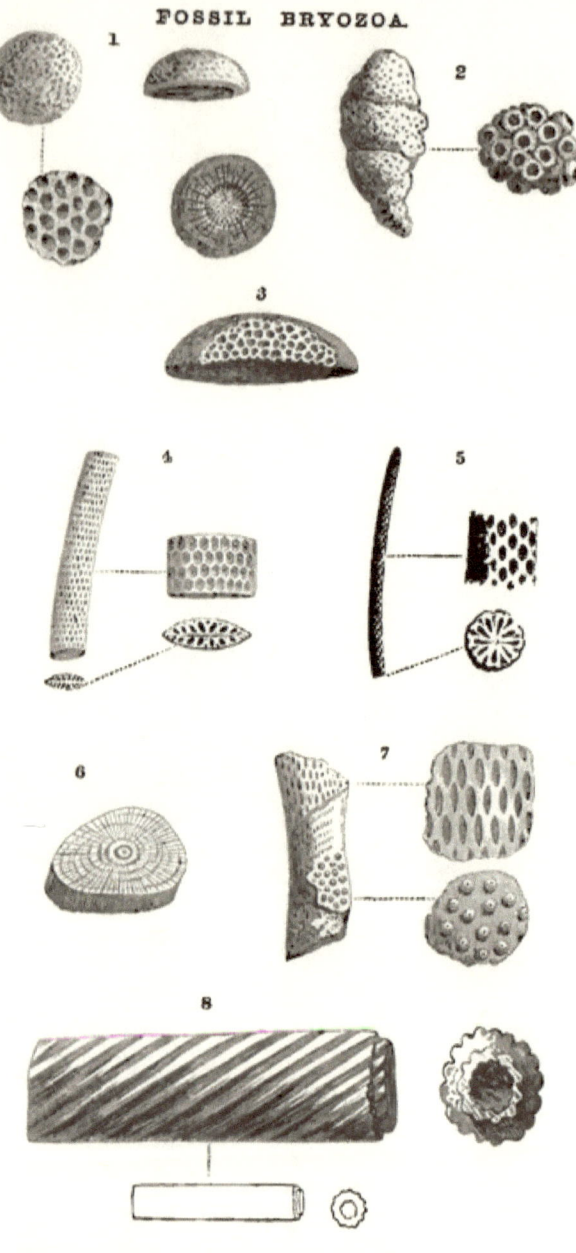

1. *Cellepora nummularia.* Busk MS. Mt. Gambier, common.
2. *Cellepora spongiosa.* „ „ „
3. *Cellepora hemispherica.* „ „ very common.
4. *Melicerita angustiloba.* Busk. „ common.
5. *Salicornaria sinuosa.* A. Hassall. „ very common.
6. Joint of stony axis of *Iris*.
7. *Eschara.* Very common throughout the whole district.
8. Axis of Coral, showing growth by deposition of calcareous matter from outside.

remote, that 'the twilight of fable' is but as yesterday in comparison.

Besides the *Foraminifera*, there is also to be mentioned the minute shells of *Entomostraca Brachiopoda*, but, beyond observing them, no attempt was made at their classification; for the rest, the stone is a most strange mixture. It is chiefly composed of minute corals and coralline, but Bryozoas are the most numerous of all; these, being so minute, are crowded together in a very compact manner, and connected by the fine paste just spoken of.

For the information of my non-scientific readers, I must explain what *Bryozoa*, called also *Polyozoa*, means. It comes from the Greek words, βρύον, moss, and ζῶον, animal; and is used to signify those polypes which are enclosed in small calcareous or horny sheaths (moss corals), which they sometimes also invest. They differ from true coral in having a more complex organisation, and, though much smaller, the rudiments of a digestive apparatus and nervous system have been discovered in some of them. They also generally possess small vibratile cilia on their tentacles, in all of which particulars they are superior to their larger brethren, the true coral. Besides, they are common to all latitudes, while the latter are uncommon outside the tropics. It is necessary to bear in mind the distinction between the *Bryozoa* and true Corals, because it will become evident, as we go on, that there has been an extensive growth in southern Australian seas of the former; and, as this is a peculiar case, the fact of its differing from a tropical coral reef, in the nature of the animals which built it, will help to explain any anomalies which may arise.

In addition, there is the true coral—there are many shells, generally small, and more commonly univalves than bivalves. It would be difficult for language to give an idea of how the fossils are associated together. Sometimes the shells are whole, but more frequently only casts, but the *Bryozoa* are generally intact, and preserved just as they grew. Occasionally, a mere cast of some bivalve shell is found encrusted all over with *Flustradæ*, and then the cast itself is entirely composed of small *Bryozoa*, so 'felted' together that it seems like one fossil, comprising within itself the features of many distinct families. Thus a *Pecten* will abruptly terminate in a *Retepora*, and

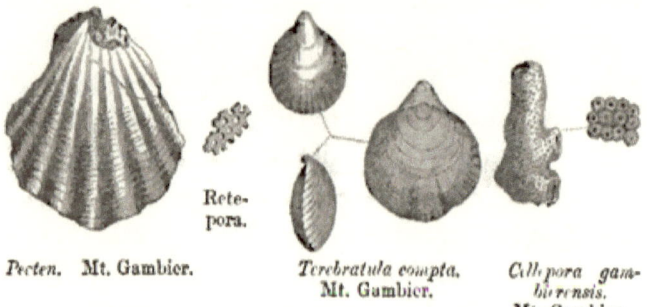

Pecten. Mt. Gambier.   Retepora.   Terebratula compta. Mt. Gambier.   Cellepora gambierensis. Mt. Gambier.

this, again, will pass into the calcareous axis of a *Pennatula*, which is stopped before it has time to display its fair proportions by the upper valves of a *Terebratula*, this being dovetailed into a mass of *Celleporidæ*, and the small spires of the *Spatangus*, which everywhere abound. There are no Nummulites, or any signs of fossils connected in any way with secondary rocks.

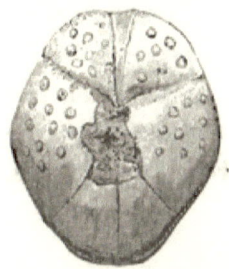 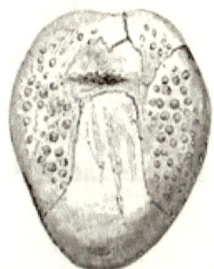

Spatangus Forbesii.   Mount Gambier.

There are, however, some singular fossils found in the above portion of the district that are worth mentioning.

Some rocks, not far from Penola, are entirely composed of a small shell not unlike *Nummulina*, but differing in many respects: they are being considered by persons more competent than myself. This is the only part of the district where they occur. I may state also, that two Lunulites were found—they were both microscopic. The organic remains are not all equally abundant in the same strata; some prevail more in the lower, while others are more common in the upper beds.

In a place near Mount Gambier, where the falling in of a subterranean hollow (probably eroded by water) has given rise to a deep circular pit, about 100 feet wide and 90 deep, a complete section of the strata is exposed as far as the depth goes. It is here seen that, in addition to a distinct line of stratification, dividing the rock into layers about fourteen feet thick, there are regular zones where particular fossils are associated. Thus at the first bed (fourteen feet), little is seen but *Bryozoa* and *Terebratulæ*; in ten feet next, less of the moss corals, and more pectens; the next is almost exclusively composed of a pecten common to this formation, with imbricated striæ, called Pecten Coarctatus, and a cellepore coral subse-

Pecten coarctatus (?).
Mt. Gambier.

Cidaris.
Mt. Gambier.

quently to be described. This state of things is nearly continued to the bottom, where Echini and Reteporæ combine with the general mass. In all the strata, the shells, &c., are cemented together. It is not contended that this arrangement is found throughout the district; but fossils are found in much the same way at the Mosquito Plains caves (seventy miles distant), where a fine section is exposed to view, and therefore it would seem that the distribution is pretty uniform. A Table of all the fossils hitherto discovered by myself is here appended; but it must be remarked, that the list is far from being intended as complete. Very likely it is but an infinitesimal fragment of what remains to be brought to light; and when it is remembered that all our knowledge of the fauna is derived from some twenty caves, and about five times as many wells that have been sunk in different places, who

FOSSILS OF THE FORMATION.

can calculate what remains yet to be discovered? In the following list the arrangement adopted by Professor Phillips, in his excellent 'Manual of Geology' (*Encyclopædia Metropolitana*) has been adhered to:—

ORGANIC REMAINS.

PLANTS . . . . . . none.

FORAMINIFERA.

Many genera, families, and species, some probably new.

ZOOPHYTA.

	No. of Species.
Isis	1
Corals, not classified	7

ECHINOIDEA.

Cardiaster	1
Cidaris	2
Echinolampus	2
Clypeaster	1
Spatangus	3
Echinus	1

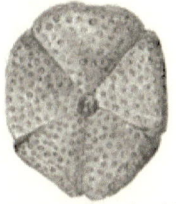

Clypeaster. Mt. Gambier.

Cast of *Trochus.*

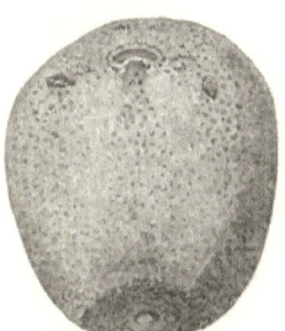

Echinolampus. Mt. Gambier.

FOSSILS OF THE FORMATION.

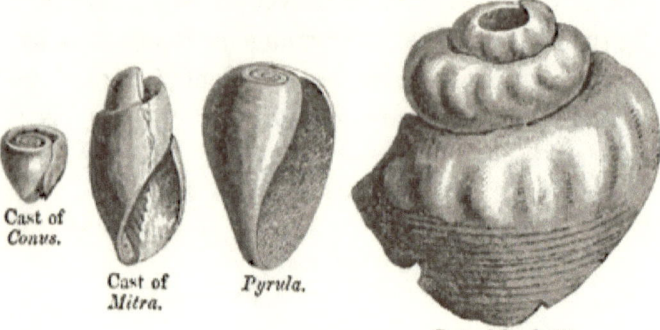

Cast of *Conus*.
Cast of *Mitra*.
Pyrula.
Cast of *Turbo* (?).

ASTEROIDEA.

	No. of Species.
Astropecten	1

CIRRIPEDIA.

Balanus	1

ENTOMOSTRACA.

Many species observed of Cypris, Cythera, &c.

BRYOZOA.*

Salicornaria	2
Canda	1
Onchopora	1
Membranipora	4
Lepralia	4
Cellepora	6
Eschara	8
Retepora	1
*Paileschara	2
*Cœleschara	1
Melicerita	1
Sertularia	1
Pustulipora	1
Idmonea	2
Hornera	2

and numerous others, belonging probably to entirely new genera; but, as these fossils are the prevailing

* Those names marked with an asterisk are new genera.

ones of the formation, the list is probably inexhaustible.

BRACHIOPODA.

	No. of Species
Terebratula	4

MONOMYARIA.

Lima	1
Ostrea	1
Pecten	5
Pinna	1

DIMYARIA.

Arca	1
Astarte	1
Cardium	2
Pectunculus	1
Panopæa	1
Cyrena	1
Venus	1

GASTEROPODA.

Ancillaria	1
Buccinum	1
Bulla	1
Cerithium	1
Conus	2
Dentalium	1
Cypræa	3
Fasciolaria	1
Fissurella	1
Fusus	1
Hyponyx	1
Littorina	1
Melania	1
Mitra	1
Murex	2
Nerita	1
Oliva	1
Pyramidella	1
Pyrula	2
Trochus	many species.
Turbo	many species.

		No. of Species
Turritella	. . . .	many species.
Voluta .	. . . .	many species.

CEPHALOPODA.

| Nautilus | . . . . . | 2 |

PISCES.

Many teeth, the most common of which appear to belong to a species of *Oxyrrhinus*, which Prof.

Teeth of Shark. *Oxyrrhinus Woodsii.*
(M'Coy, M.S.) Mt. Gambier.
These are natural size, but many are found four times larger.

M'Coy has called *Oxyrrhinus Woodsii*. Some of them seem to be those of a Lamna. *No* fish-bones were found.

Aves, Mammalia, Marsupialia, Insectivora, Cheiroptera, and *Quadrumana*, have none of them, if existing in this district when the beds were deposited, left any traces that have yet come to light. Of *Cetacea*, some bones have been found at various times, but have never been examined by the author. The Murray banks seem to be the commonest locality for these remains. With regard to *Marsupialia*, although instances will be hereafter given of such bones being discovered, they have never been deposited in the limestone and associated with marine remains. The instances alluded to are generally in connection with caves.

In addition to the list just given, fossils were found which would not properly come under any of the above headings; such, for instance, as extensive beds of spines of *Echinidæ*, amongst which the spines of a *Cidaris* are so large and highly tuberculated as to seem like distinct shells, while the form of others is so peculiar as to earn for them the ludicrous title of 'fossil cribbage-pegs,' which, indeed, they are not unlike.

Spine of *Cidaris*. Mount Gambier.

Some remains of crabs' claws, which are not unfrequently met with, have not been included in the above list, simply because they were so imperfect as to render any classification very difficult and uncertain. Long strips of hardened lime, which probably belonged to the calcareous ages of *Pennatulidæ*, are also common. Some fragments of seaweed were said to have been found in the northern part of the district, but the specimens were lost, and their true nature is consequently very doubtful.

In nearly every case, the lime in the shelly portions of the fossils is crystallised, and fractures always in the crystalline form, leaving a smooth even edge. The lime inside is never crystallised. This phenomenon, which is common to many formations in Europe, is worthy of more consideration than it has received. Can it be that the decomposition in the animal matter in the shell causes a chemical arrangement of the other particles connected with it, by a sort of predisposing affinity? Instances like this are not uncommon in chemistry.

Before comparing this list with the fauna of other parts of this and the neighbouring colony, where fossils are found, let us stop to examine the nature of the evidence they furnish. In the first place, the beds are tertiary. This is seen from the fact that some of the fossils are of existing species, and from the general resemblance of the fauna to what is now found on the coast. Secondly, the formations are most probably of the Lower Crag or Middle Crag; but this will require some little explanation. Most readers are probably aware that the tertiary or newest fossiliferous rocks have been divided into three great periods. The Pleiocene is most recent in the species of shells it possesses; the Meiocene is less recent; and the Eocene, or dawn of the recent, the earliest. The grounds of this division are found, as the names imply, in a greater or less predominance of existing forms, among the fossils enclosed. In the beds now under consideration, there can be no question that a number of the shells still exist on the coast, but not by any means in proportion to the past; Pleiocene is sometimes found above them, in which extinct species are rare. On comparing the fauna in the few instances where any likeness exists with shells found in Europe, they are found to be most commonly similar to Lower Meiocene and

Upper Eocene fossils. Thus, the *Nautilus ziczac* is found in Upper Eocene at home, but sometimes

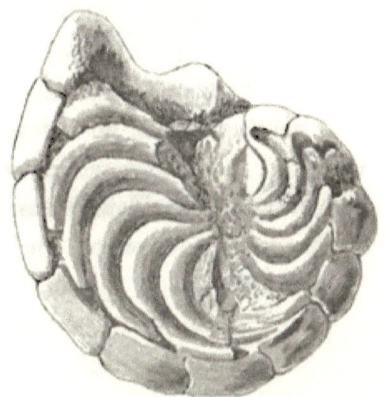

Nautilus ziczac. Mount Gambier.

goes as low as the London Clay, in which are also found the *Astro-Pecten*, the *Spatangus Forbesii*, and

Spatangus Forbesii. Mount Gambier.   Cast of *Turritella terebralis.* Mount Gambier.   *Murex asper.* Lam. Mount Gambier.

the *Cypræa oviformis*, common to this formation. The most predominant shell found at Mount Gambier is the *Nautilus ziczac*, and on the Murray the commonest shell is the *Turritella terebralis*, common in the Meiocene beds, Bordeaux. Of course, in dealing with any tertiary rocks, it is difficult to say what shells have become extinct, and what may yet be found in future exploration, so that those which are common in the Eocene beds might just possibly be found in Pleiocene beds, and thus deposits so widely separated be found similar in some respects. As an instance, it may be mentioned that the *Murex asper*, Lamark (Upper Eocene, Barton clay, Hants), has been found by myself, at least, as a common shell on some parts of the Australian coast. Now, Prof. Rupert Jones informs me that the microscopic fossils are not indicative of anything more ancient than Pleiocene. If this be the case, we must regard the occurrence of the fossils enumerated above as continuations of animal life here, which were extinct elsewhere. Some deductions will hereafter be made from the general nature of the strata, and the occurrence of Eocene fossils in Pleiocene rocks, while some, even from a portion of our present fauna, are singularly corroborative of the view taken. But this is anticipating.

The following report of Dr. Busk on some fossils of the formation seen by him, will show how far the position of the beds has been determined as yet:—

'The *Polyzoa* included in this collection belong to fifteen or sixteen genera, of which four are probably new; and the number of species is about thirty-nine or forty, of which at least thirty-six seem to be undescribed. Among them are several very peculiar and characteristic forms, especially in the genus *Cellepora*. Taken as a whole, these fossil forms exhibit such genuine and specific types as to render it probable that the formation in which they are found corresponds, in point of relation, to the existing state of things with the lower crag of England, although the collection contains only one or two species, and that even doubtfully, to any belonging to the crag.

'It is remarkable, however, that it presents a second species of *Melicerita*, which genus is peculiar to that deposit. Of the characteristic *Fasciculariæ* and other *Theonidæ* of the crag, no trace exists in the present collection. The most remarkable form

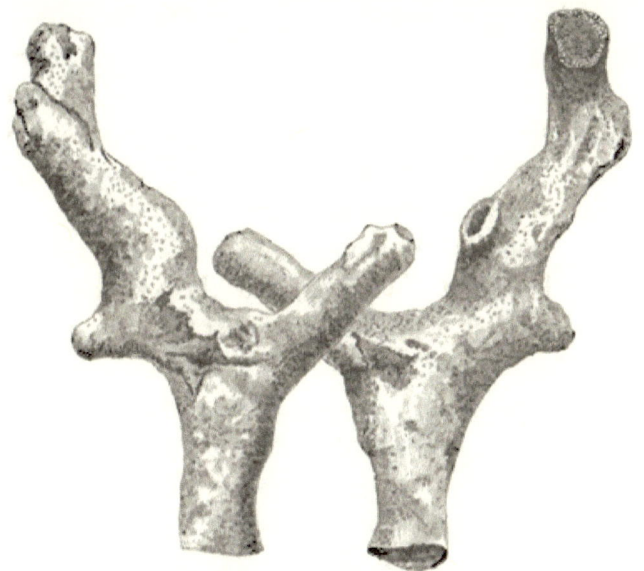

Cellepora gambierensis. Busk.
Mount Gambier.

is a large and massive *Cellepora*, for which I propose the name *Cellepora gambierensis*.'

It appears to me almost certain that eventually future investigations will identify these deposits with the Crag, and therefore it will be as well here to mention the features of that deposit at home.

The word 'crag' is supposed to be derived from an ancient British word, meaning rock, and the term is therefore a provincial one. The beds are best known by those in Suffolk, where they are divided into the upper and lower crag. The upper is a loose shelly deposit, seeming like a shifting sandbank, whose description is exactly similar to a deposit to be noticed by and by, which immediately overlies a portion of the Mount Gambier limestone; the lower is a soft coralline limestone, precisely similar to what I have described as the rock at Mount Gambier. I believe this deposit and the chalk have always been regarded as the two which are richest in Bryozoa, and wherever it occurs it has merited the name of coralline limestone, from its peculiar richness in those remains. It has been remarked, also, that these remains indicate a peculiar state of the sea at the time, which is not accounted for by anything we observe at present going on about us. When this is remembered in connection with the peculiar origin of the beds of which I am now about to treat, it will be admitted that there is a similarity, even if no shells were there — that is, noshells common to both.

The crag deposits have been traced at home, to Antwerp, Normandy, the Apennines, and to many parts of Italy, not even excepting Rome. When these beds are proved to be identical, as doubtless they will shortly be, by the discovery of many similar fossils, it will be established that the sea which formed the strata upon which Mount Gambier, and even Adelaide, stand, was also rolling over the site of Rome—not ancient Rome, but Rome then unborn. When we look back on the history of that country, and think of the period before its upheaval from the sea, we can guess how *modern* these beds of Mount Gambier are. That the sea covered both simultaneously does not admit of much doubt.

I am almost ashamed to quote so largely in a small work like the present, but, as illustrative of what I have been describing, I cannot help transcribing from Sir Charles Lyell's invaluable manual the passage referring to the Pleiocene strata of Rome:—

'The seven hills of Rome are composed partly of marine tertiary strata; those of Monte Mario, for example, of the older Pleiocene period, and partly of superimposed volcanic tuff, on the top of which are usually cappings of a fluviatile and lacustrine deposit. Thus on Mount Aventine, the Vatican, and the Capitol, we find beds of calcareous tufa, with incrusted reeds and recent terrestrial shells, at the height of about 200 feet above the alluvial plain of the Tiber. The tusk of the mammoth has been procured from this formation, but the shells appear to be all of living species, and must have been embedded when the summit of the Capitol was a marsh, and constituted one of the lowest hollows of the country as it then existed. It is not without interest that we thus discover the extremely recent date of a geological event which preceded an historical era so remote as the building of Rome.'*

The deposit is further traced into Asia Minor and on the shores of the Caspian Sea. It is certainly very interesting to find in what manner Australia is geologically connected with the older hemisphere, and especially to find that we are related by very close ties of time with such distinguished ancients as Rome and Asia Minor. There is one important difficulty in tracing the resemblance, which has yet, in great part, to be overcome. Probably not a dozen fossils will be found common to both formations; but, inasmuch as the fauna of our present seas differ almost totally from those of Europe, so we must expect a similar divergence for the time when the crag was deposited. The problem to be solved will therefore stand thus: Knowing the present analogies between the Australian and European seas, what might we anticipate from the Australian crag, knowing the fauna of that of Europe?

It might be further remarked, that the discovery of extensive beds of coralline, all belonging to one period, might enable us to establish an epoch of Bryozoa, just as there is an epoch of carbonaceous flora, an epoch of gigantic reptiles, or an anomalous period like the chalk, which seem, each in their turn, to have stamped a character on every part of the world during their continuance.

* Lyell's *Manual of Geology*, 5th ed., p. 176.

From the general appearance of the strata, it may be concluded that they were deposited in a deep tranquil sea; that the débris of which they were composed was derived from a series of coralline reefs, which either at present form a part of them, or somewhere in their vicinity; and, thirdly, that the climate was somewhat warmer than that which obtains at the present time in the same area. The inference with regard to the depth of the sea has been drawn from the nature of the fossils, which are generally indicative of considerable depths. At least, it will be proved just now that the animals did not exist where their fossil remains are now found. Then there are no river or land shells, or bones of mammalia, or trees, or wood, such as we might expect if land were near. Very few of the Zoophytes belonging to the fossils appear to have lived and died in the places where their remains are found. It would appear as if they had been brought from a distance, and that the carriage has been effected not so much by the force of a current as by the gradual spreading out upon the bottom of the ocean, either by the force of gravity, or the pressure of detritus from the shallower sea. It is true that, as will be hereafter shown, corallines (very similar to corals) are sometimes found in the position in which they have grown, proving the comparative shallowness of the sea, which was perhaps not more than thirty fathoms, which is the greatest depth at which live corals have been found. Yet such instances are not common, and it is very probable that the detritus from these and similar places have given rise to the marine exuviæ now embedded in the stone. In such places, again, the shells are nearly always preserved, are more numerous, and the beds contain less of that limestone which unite the other rocks; while elsewhere the shells are fewer, more broken, and oftener only casts remain.

The regularity of the strata, and the dim traces of stratification, point to a very tranquil sea, which, of course, could only be obtained at some depth; and where portions of shells are found, or pieces of fish-bone, teeth, &c., the remaining fragments are seldom discovered in the immediate vicinity, thus proving a transit from the original place of deposition which has separated the remains tranquilly and without much breakage. If I were asked to indicate localities which would be good types of where the process of slow shifting has taken place, I should point out the rock about Mount Gambier, where, of all the fossils found, not one animal appears to have died in the place where the remains are now found. The coral before alluded to is very common here, but always broken, and appearing to have come from a distance. On the other hand, the only place I am as yet acquainted with where the fossils do not appear to have been transported any distance, is at the cliffs in the caves at Mosquito Plains.

The mention of these brings me to the proof of the third statement, namely, that the whole of the formation is derived from a series of coralline reefs, perhaps now forming part of the beds. In the first place, the most common fossil in the whole formation, whether at the extreme south or far north of the district, is the *Cellepora gambierenssi*,

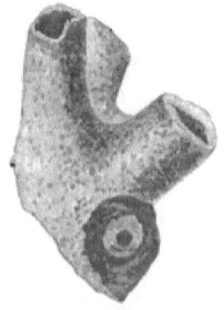

Branching Axis of *Cellepora gambierensis*.
Mount Gambier.

before mentioned. It is found at all depths and in every place where the beds occur, at Mount Gambier, in the caves at Mosquito Plains, at the edge of the Mallee Scrub, and in the Murray Cliffs. Most frequently it is in small pieces, but occasionally large and branched. Often it is so worn and broken as to be barely traceable in the limestone cement, and I have noticed it even in the centre of flints. At the caves, Mosquito Plains, it appears as if it had grown where it is seen in the walls. It is very large and much ramified; not, certainly, possessing the beauty of the delicate *Mæandrina* (brain coral), or the leaf-like expansion of the *Pavonia*, but, nevertheless, having a natural beauty of its own, all the better for a close examination, where evidence is obtained of the minute kind of molluscous animal which inhabited it. This is the only coralline which seems to affect the large massive branches of the true corals, forming a connecting link with that order. It is made up of closely-packed congeries of minute cells, which opened outward, in pores, as seen in the engraving in this chapter, where a magnified portion of the surface is given. The cells seem to have grown from within outwards, and the inside of the branches is hollow. A section is much like very fine yellow sponge. It might easily have formed a reef.

Besides the actual presence of the coralline to point out the origin of the strata, their very texture, if the term be admissible, proves the nature of the process which formed them, which appears from the following observations.

Lieutenant Nelson * states that the mud derived from coral reefs differs not, when dried, from the ordinary white chalk of Europe, and 'this mud is carried to great distances by currents, and spread far and wide over the floor of the ocean.'

It seems quite natural to suppose that this would be the case, from the heavy beating of the open sea upon coral reefs. This not only breaks up the coral, but carries it far away, in the form of white mud, with small fragments of coralline shells, &c., interspersed. In the lagoons found in the centre of Atolls, or Ring Islands, such as abound in the

* *Quarterly Journal of Geological Society*, 1853, p. 200.

Pacific, the same description of white mud is met with; that of Keeling Island is thus described by Mr. Darwin: *—

'The sediment from the deepest parts of the lagoon, when wet, appears chalky, but, when dry, like very fine sand. Large soft banks of similar but even finer-grained mud occur on the SE. shore of the lagoon, affording a thick growth of a *Fucus*, on which turtle feed. This mud, although discoloured by vegetable matter, appears, from its entire solution in acids, to be purely calcareous. I have seen in the Museum of the Geological Society a similar but more remarkable substance brought by Lieutenant Nelson from the reefs of Bermuda, which, when shown to several experienced geologists, was mistaken by them for true chalk. On the outside of the reef much sediment must be formed by the action of the surf on the rolled fragments of coral, but in the calm waters of the lagoon this can take place only in a small degree. There are, however, other and unexpected agents at work here: large shoals of two species of *Icarus*, one inhabiting the surf outside the reef, and the other the lagoon, subsist entirely, as I was assured by Mr. Liesk, the intelligent resident before referred to, by browsing the living polypifers. I opened several of these fish, which are very nume-

---

* I cannot let this opportunity pass without expressing my great obligations to Mr. Darwin's valuable work on coral reefs. Its merits cannot be too highly estimated, and to any one studying the geology of such deposits as those of Mount Gambier, it is an indispensable text-book.

rous and of considerable size, and I found their intestines distended by small pieces of coral and finely-ground calcareous matter. This must daily pass from them as the finest sediment; much also must be produced by the infinitely numerous vermiform and molluscous animals which make cavities in almost every block of coral.'

As we are upon this subject, it is hoped it will not be deemed tedious to insert, for the information of those unacquainted with coral reefs, a description of one, a proper notion of them being indispensable to a comprehension of this chapter.

The following is taken from the voyage of H.M.S. *Fly* to the Eastern Archipelago.* A coral reef is thus described:—

'A submarine mound of rock, composed of the fragments and detritus of corals and shells compacted together into a soft spongy stone. — The greater part of the surface of this mound is quite flat, and near the level of low water. At its edges it is commonly a little rounded off, or slopes gradually down to a depth of two, three, and four fathoms, and then pitches suddenly down, with a very rapid slope, into deep water 20 or 200 fathoms, as the case may be. The surface of this reef, when exposed, looks like a great flat of sandstone with a few loose slabs lying about, or here and there an accumulation of dead broken coral branches, or a bank of dazzling white sand. It is, however, chequered with holes and hollows more or less deep, in which small

* London: Boone, 1847.

living corals are growing, or has, perhaps, a large portion that is always covered by two or three feet of water at the lowest tides, and here are fields of corals, either clumps of branching Madrepores, or round stools and blocks of *Mœandrina* and *Astræa*, both dead and living. Proceeding from this central flat towards the edge, living corals become more and more abundant.

'As we get towards the windward side, we of course encounter the surf of breakers long before we can reach the extreme verge of the reef, and among these breakers we see immense blocks, often two or three yards (and sometimes much more) in diameter, lying loose upon the reef. These are sometimes within reach by a little wading, and though, in some instances, they are found to consist of several kinds of corals matted together, they are more often found to be large individual masses of species which are either not found elsewhere, and, consequently, never seen alive, or which greatly surpass their brethren on other parts of the reef in size and importance. If we approach the lee edge of the reef, either by walking or in a boat, we find it covered with living corals, commonly *Mœandrina*, *Astræa*, and *Madrepora*, in about equal abundance, all glowing with rich colours, bristling with branches, or studded with great knobs and blocks.

'When the edge of the reef is very steep, it has sometimes overhanging ledges, and is generally indented by narrow winding channels and deep holes leading into dark hollows and cavities, where nothing can be seen. When the slope is more gentle, the great groups of living corals and intervening spaces of white sand can be still discerned through the clear water to a depth of forty or fifty feet, beyond which the water recovers its usual deep blue. A coral reef, therefore, is a mass of brute matter, living only at its outer surface, and chiefly on its lateral slopes.'

This description must be varied for the majority of atolls. The outer edge is a bank of calcareous sand, raised a few feet above the highest tides. This is sometimes covered with a few shrubs and palm trees. Inside the bank of sand which forms a ring there is a lagoon, the bottom of which is covered with a chalky mud just spoken of. It has been very nearly proved that the chalk of England and France owes its origin to a somewhat similar process; and from the similarity of the strata in both cases (for really the cliffs at Mount Gambier can scarcely be more like the chalk cliffs), all the arguments which tend to support such a view will serve equally well here.

They principally rest on the fact, that, where coral reefs are at present in existence, a deposit analogous to the chalk is in course of formation, and that where chalk beds are found it is always associated with such fossils and exuviæ as might be looked for from a coral reef.

Let it be remarked, that the quantity of true corals which have been found in the Mount Gambier strata is admitted to be comparatively small, yet the same thing might be urged against the chalk at home.

There, as here, the *Bryozoa* are in excess, and probably there, as here, the latter formed the reefs. Such a thing is never observed now, but I am confident it must have been the case here in the geological period to which they belong. As just related, there is a fine section of the beds at the caves at Mosquito Plains.

There is little else in that place but the *Cellepora gambierensis*, common to the formation. It stands exactly as it grew, large branches dividing again and again, as they ascend, until at the top of the cave their arborescent ramifications are spread out like a shrubbery turned to stone. Now, although the *Cellepora* is not a true coral, it might easily form part of a reef when growing as related, and there cannot be much doubt that this particular instance was in actual connection with one. It may therefore be enumerated, that there may be reefs of 'moss coral,' *Bryozoa*, as well as of true coral, and the Mount Gambier deposit was derived, probably, from the former.

What other fauna may we expect besides corals and moss corals in a reef? If we consult the works of those who have made such places their study, we shall find that *Pectens, Ostreæ, Conidæ, Cypridæ, Sharks, Echini,* and *Bryozoa* generally, are everywhere associated with reefs. For these facts, see Darwin, Beechy, Chamisso, Moresby and Jukes, *passim*. Readers now have only to refer back to the part of this chapter which mentions the fossils of the beds, and they will at once see the correspondence with the faunæ of a coral reef. It may, then, be stated, without taking anything for granted, that the deposit in question has arisen from a reef, or series of reefs, now, perhaps, entombed in the general débris, perhaps in some parts of the strata not yet known.

It will be necessary to postpone any enquiries as to the nature of the reef — whether a barrier or a series of atolls, or what other kind or kinds of reefs. The marked boundaries of the formation have been described, at least, as far as they are at present known. In the mean time, we may pass to the consideration of another fact, before spoken of, namely, that the organic remains indicate a much warmer climate than that which obtains at present. Some little explanation may be further necessary as to the natural history of the builders of the reefs, to prevent any misconception of what is meant to be conveyed by the term 'coral.'

It is well known that true coral reefs at present do not extend much beyond the tropics, either in a northerly or southerly direction, except at the Bermudas, where local circumstances of temperature (the Gulf Stream) account for the exception. It is not pretended, even now, that there was any variation from the law during the crag period. A coral reef, properly speaking, is a mass of calcareous polypodoms, formed by living polypi. Such a definition would exclude our reef, at least as far as the fossils are known, because the animal which raised the barrier was of a much higher organisation than a polypus or real coral animal, and such an one as is even now found outside the tropics. It is a molluscous animal, which is most common in our beds, belonging to the sub-kingdom *Nematoneura*, and associated with such divisions as *Echinodermata*, such as star-fish, &c., whereas the true corals belong to the sub-kingdom *Acrita*, and are associated with the *Porifera* or sponges, the *Acalephæ* or jellyfish, &c. The higher developement of the *Bryozoa* might, perhaps, enable them to withstand a colder temperature. At any rate, this is not the only instance of fossils indicating something analogous to a coral reef being found outside the tropics. The chalk formation extends beyond the 55th parallel of north latitude, while, in the case under consideration, it does not certainly extend beyond 42° south. I do not believe, however, that the same animal would exist now, even at the latitude of Mount Gambier* (about 37° 30′ south), and, therefore, a warmer temperature must be supposed than that which is consistent with the present physical geography of Australia. This is borne out by the fauna. *Conidæ*, *Cypridæ*, and *Nautilidæ* are not found now of the same size upon the coast, except much nearer to the tropics.

And now we come to the important question, How far does the formation extend? It may be

---

* My reasons for this belief are taken from the present fauna of that part of the coast.

necessary to go somewhat extremely into detail for a proof of what is certainly a very extraordinary fact, probably not a solitary one, in the geological history of the tertiary formation, and although this work only professes to record geological observations in South Australia, yet, in this instance, it will be led somewhat beyond these limits.

It may be premised that a large area, covered by one formation, is by no means new in geology. When it is stated that probably one-sixth of Australia is covered by the one now described, its extent may surprise us. The chalk formation of Europe has, however, been traced 1,140 geographical miles in one direction, and 840 miles in another. The one we are now occupied with is not so large as this, though it is of great size; but, as many of the facts in support of this view do not amount to decisive evidence, their separate value must be shown as we proceed step by step;—this will be the subject of the next chapter. Before passing to it, let us consider for one moment the magnitude of the operations we have just cursorily surveyed. The sight of an atoll standing alone amid the surgings of a vast ocean, with no other guarantee against being swept away by some great billow than the protection of the tiny world which raised the structure, may well surprise us ; but here the labours of a similar creature are as marvellous. An immense territory, much larger than Great Britain, owes its soil, its foundation, and hereafter will owe its edifices and monuments, to fragile corallines, of which we tread a million under our feet in our sea-side rambles, and whose only appeal against destruction is being so very small and yet so very beautiful. They raised a structure which withstood the waves, and now—

'Pulvere vix tectæ poterunt monstrare ruinæ.'—LUCAN.

It seems as though the order of nature were reversed, and that weakness had power to complete what strength had never power to attempt; that while the monuments of giants perish, the gigantic monuments raised by atoms remain for ever. Well may we here quote the beautiful lines from Montgomery's 'Pelican Island:'—

'I saw the living pile ascend,
The mausoleum of its architects:
Still sloping upwards as their labours closed—
Slime the materials, but the slime was turned
To adamant by their petrific touch.
Frail were their frames; ephemeral their lives;
Their masonry imperishable. All
Life's needful functions, food, exertion, rest,
By nice economy of Providence,
Were overruled to carry on the process,
Which out of water brought forth solid rock.
Atom by atom, thus the burden grew—
A Coral Island, stretching east and west.
Steep were the flanks, with precipices sharp,
Descending to their base in ocean gloom.
Chasms, few, and narrow, and irregular,
Formed harbours safe at once and perilous—
Safe for defence, but perilous to enter;
A sea-lake shone, amidst the fossil isle,
Reflecting in a ring its cliffs and caverns,
With heaven itself, seen like a lake below.
Compared with this amazing edifice,
Raised by the feeblest creatures in existence,

What are the works of intellectual man,
Towers, temples, palaces, and sepulchres?
Dust in the balance, atoms in the gale,
Compared with these achievements in the deep,
Were all the monuments of olden time.
Egypt's grey piles of hieroglyphic grandeur,
That have survived the language which they speak,
Preserving its dead emblems to the eye,
Yet hiding from the mind what these reveal —
Her pyramids would be mere pinnacles,
Her giant statues, wrought from rocks of granite,
But puny ornaments for such a pile
As this stupendous mound of catacombs,
Filled with dry mummies of the builder-worms!'

CHAPTER V.

AN UNFINISHED CONTINENT.

EXTENT OF THE FORMATION. — MURRAY CLIFFS. — STURT'S LIST OF FOSSILS. — DESCRIPTION OF THE CLIFFS. — EXTENT OF THE FORMATION IN A WESTERLY DIRECTION. — STURT'S ACCOUNT OF THE FORMATION TO THE NORTH. — FLINDERS' DESCRIPTION OF THE SOUTH. — OTHER OBSERVATIONS. — BOUNDARIES TO THE EASTWARD. — TASMANIA. — ORIGIN OF THE FORMATION. — SHOWING SUBSIDENCE OF A LARGE AREA. — DARWIN'S THEORY. — APPLICATION OF THIS TO THE MOUNT GAMBIER BEDS. — OBJECTIONS ANSWERED. — WHY NO REMAINS OF ATOLLS ARE FOUND. — PROBABLY SOME REMAINS AT SWEDE'S FLAT. — PROBABLE TEMPERATURE OF THE SEA. — GEOLOGICAL PERIOD. — ANALOGIES IN THE PRESENT STATE OF THE EARTH'S CRUST WITH FORMER GEOLOGICAL EPOCHS. — ANALOGY OF AUSTRALIA TO THE CHALK. — RETARDED STATE OF ITS ZOOLOGY. — BAD ADAPTABILITY AS A RESIDENCE FOR MAN. — CONCLUDING REMARKS.

IT will be remembered that the district to which this book refers is bounded on the north and west by the river Murray, and on the west to the south by the sea, containing an area of about 22,000 square miles. Now, all this immense tract of land is, as before stated, occupied by the same formation, with one or two exceptions. Of this, as far as the Mallee Scrub, and even somewhat beyond it, we have proof positive. Wherever cliffs are seen or wells sunk, the characteristic shells and corals appear. A large space, of which we know little, intervenes between the commencement of the scrub and the river, and then fossiliferous cliffs of yellowish limestone line each side of the stream. As there are no elevations of any importance known in the scrub, and as the soil seems to preserve the same character throughout which is seen in the commencement, it is no very great stretch of hypothesis to believe that the formation is continuous, that is, if the cliffs on the Murray are of the same nature. But there can be but little doubt of this.

When Captain Sturt first traced down the Murray from the Murrumbidgee in 1829, he came suddenly to a part of the river somewhat eastward of the present boundary of South Australia, where it was bounded on each side by high limestone cliffs, through which the stream seemed to have worn a bed. These continued right down to the sea mouth, and even then seemed to be prolonged along the coast to the south-east. By referring to a map, readers will see that the river, amid various windings, generally preserves a south and westerly course, until where it suddenly takes a bend, and then continues south to the sea. This direction of the river is just such as must cut through any intervening strata of a different nature, if there were any. But this is not the case. On the contrary, the whole seems to be of the same description of rock. From the following list of fossils, engraved or described by Captain Sturt, the identity of the fauna will at once be seen when compared with the list already given.

The catalogue was compiled in 1832; of course due allowance must be made for the nomenclature, in consequence. It may be remarked, that the plates in the work in which this list occurs show some of the species to be identical with some in my possession. Thus, what he terms the *Spatangus Hoffmanni* is the *S. Forbesii*; the *Glauconome* is the *Salicornaria*; the *Eschara celleporacea* is the *Cellepora gambierensis*, common to the whole formation. But, to facilitate the comparison, those fossils found at Mount Gambier are marked with an asterisk:—

TUNICATA.

(The classification is left as it stands in Sturt's work.)

* Eschara celleporacea.
* „ pisiformis.
  „ unnamed.
* Cellepora echinata.
  „ escharoïdes.
* Retepora disticha.
  „ silicata.
* Glauconome rhombifera.

All tertiary in Westphalia and England.

RADIATA.

Scutella.
* Spatangus Hoffmanni *Goldfuss*.
    Tertiary in Westphalia.
*Echinus.

CONCHIFERA: BIVALVED SHELLS.

Corbula gallica, Paris Basin, tertiary.
Tellina.
Corbis lamellosa, tertiary, Paris.

Lucina.
* Venus (Cytherea) lævigata, Paris.
    „   obliqua, ibid.
Venus.
Cardium? fragments.
Nucula, such as are found in London clay.
* Pecten coarctatus, Placentia.
*   „   various, recent.
     „   species unknown.
        Two other Pectens also occur.
* Ostrea elongata, *Deshayes*.
* Terebratula.
One cast, genus unknown, perhaps a Cardium.

MOLLUSCA: UNIVALVED SHELLS.

Bulla.
Natica, small.
   „   large species.
Dentalium.
Trochus.
Turritella.
   „   in gyps.
Murex.
Buccinum.
Mitra.
   „   very short.
Cypræa.
Conus.

This list would present a still greater resemblance to that of Mount Gambier, were I able to avail myself of the numerous collections that have been since made of that locality. That the deposits are identical, I have no doubt. It is true that the limestone is of a yellowish colour, while at Mount Gambier it is brilliant white, but such a change takes place at the Mallee Scrub, and appears to be due to local circumstances. The same clay is seen in the soil above the Murray cliffs, which support the same flora, and is identical in all respects. Perhaps the fossils are more tropical in the Murray. They contain more *Cephalopoda*, larger *Terebratulæ*, and, in fact, generally a larger description of *Testacea*, though of the same species as those of Mount Gambier. But the difference of temperature must be taken into consideration; the latter is nearly three degrees of latitude farther from the tropics.

The cliffs are not always upon both sides of the river, but sometimes on one and sometimes on the other. This arises from the current coursing round different elevations of the rock. The formation was first met with by Captain Sturt, at about long. 140° east, and is thus described by him:—

'As we proceeded down the river, its current became weaker, and its channel somewhat deeper. Our attention was called to a remarkable change in the geology of the country, as well as to an apparent alteration in the natural productions. The cliffs of sand and clay ceased, and were succeeded by a fossil formation of the most singular description. At first, it did not exceed a foot in height above the water, but it gradually rose like an inclined plane, and in colour and in appearance resembled the skulls of men piled one upon the other. The constant rippling of the water against the rock had washed out the softer parts, and made hollows and cavities that gave the whole formation the precise appearance of a catacomb.

'On examination, we discovered it to be a compact bed of shells, composed of a common description of marine shells, from two to three inches in length, apparently a species of *Turritella*.

'At about nine miles from the commencement of this formation it rose to the height of more than 150 feet; the country became undulating, and a partial change took place in its vegetation. We stopped at an early hour to examine some cliffs, which, rising perpendicularly from the water, were different in character and substance from any we had as yet seen. They approached a dirty-yellow ochre in colour, that became brighter in hue as it rose, and, instead of being perforated, were compact and hard. The waters of the river had, however, made horizontal lines upon these fronts, which distinctly marked the rise and fall of the river, as the strength or depth of the grooves distinctly indicated the levels it generally kept. It did not appear from these lines that the floods ever rose more than four feet above the then level of the stream, or that they continued for any length of time. On breaking off pieces of the rock, we ascertained that it was composed of one solid mass of sea-shells, of various kinds, of which the species first mentioned formed the lowest part.

* * * * * *

As we proceeded down the river, we found that it was confined in a glen, whose extreme breadth was not more than half a mile.

'The hills that rose on either side of it were of pretty equal height. The alluvial flats were extremely small, and the boldest cliffs separated them from each other. The flats were lightly wooded, and were, for the most part, covered with reeds or *Polygonum*. They were not much elevated above the waters of the river, and had every appearance of being frequently inundated. At noon we pulled up to dine, upon the left bank, under some hills, which were from 200 to 250 feet in height. While the men were preparing our tea (for we had only that to boil), M'Leay and I ascended the hills. The brush was so thick upon them that we could not obtain a view of the distant interior. Their summits were covered with oyster-shells, in such abundance as entirely to preclude the idea of their having been brought to such a position by the natives. They were in every stage of petrifaction.'

At the great southerly bend very finely preserved fossils are found; and in a collection sent to me I had no difficulty in recognising (as before stated) many of the Mount Gambier species, but there were others among them which I have not as yet seen in the latter place. To the north of the river Murray the country stretches out, in an unknown plain of scrub, for very many miles, where, according to some, the same foundation may be occasionally seen. The uncertainty of such observations leaves it doubtful if such be the case, but the same flat scrub of yellow sand renders it far from improbable.

If the area covered by this one formation be

now calculated, the territory occupied by it will be seen to be immense. But we have only been considering it in its north and south direction, a little to the west of the boundary between the two colonies. Let us now follow it farther to the westward. On the map it will be seen that there is a jutting-out of the coast (Cape Jervis) to the west of the Murray mouth, which promontory is the commencement of a range of hills upon which Adelaide is placed.

The range continues in a north and south direction, with few intervals, right up to Mount Hopeless, at the bend of Lake Torrens, some 500 miles to the northward of where the range commences. It is not meant that this chain is uninterrupted, or that there are not occasionally bends-off to the north-west, and there are some in an easterly direction, as, for instance, the Barossa Ranges, and many spurs and small ranges running off at various points, but no one can look at the map without being convinced that they all belong to one chain of mountain, of which the loftiest is not more than 3,000 feet high, running in a general northerly direction.

They are all of nearly the same description of rock, namely, slates, schists, and metamorphic rocks, with occasionally granite, porphyry, and trap rocks. A description has been already given in the second chapter of this work. Let them only be borne in mind just now for the sake of getting a good idea of the geographical features of this part of Australia, in order to better understand the subject under consideration.

The fossil coral formation extends to the foot of these in a westerly direction as far as they are known, and seems to have been deposited round them. Between the chain and the river Murray there is always a large extent of flat scrubby land, known as the Murray Scrub. It possesses the usual character of such scrubs, and, wherever wells have been sunk, the usual shells have been found.

Commonly, there is a bed of oyster-shells on the top, with others of a much more recent origin than those in the strata beneath, being, in fact, mostly of existing species, and this is one of the exceptions referred to just now.

When it is stated that only one deposit is found in the whole district, it is not meant to exclude others which may overlay them. There is a newer deposit more or less distributed over this part of the country, but then the older is always underneath, and visible at a small depth, or may reasonably be presumed to be present from the fact of its dipping under the newer beds and reappearing again on the other side.

On reaching the western side of the great South Australian chain the formation disappears. On the east side, again, a little plain is bounded by the sea, which runs in an estuary to the northward of Adelaide, and bears the name of St. Vincent's Gulf. The western side of the gulf is bounded by Yorke's Peninsula, a narrow boot-shaped strip of land, in which the fossil beds appear again. High limestone cliffs bound one side of the coast, and at Kangaroo Island, which is only separated from the peninsula by a channel bearing the name of Investigator's Straits, the fossiliferous rock is repeated, at least in a few spots. To the westward of Yorke's Peninsula, a great gulf (Spencer's Gulf) runs up much farther to the north than the one just mentioned, and west of that the country stretches out in an almost unbroken scrub as far as the colony of Western Australia.

Of the nature of the geological formation we can know but little as certain; but, if we have pretty accurate notions of what lies to the north and what to the south, and know, further, that there are no ranges of any consequence intervening, we can form some idea of what rocks should exist there.

Of course such conclusions are liable to error, because, if there may be an immense difference of geological character in a small space of ground, how much more in such a space as that which lies between the points to be referred to? Of one thing we may be sure, however, and that is, that there are no mountain ranges of any altitude. If such existed, they would give rise to rivers, and in the great Australian Bight, to the south, no such features occur.

When Captain Sturt, in 1845, pushed far into the north-west interior and crossed the Stony Desert,* he found that fossiliferous limestone cropped

---

* See *Expedition to Central Australia*, vol. i.

out on each side of Lake Torrens, and, in fact, in all the flats of the interior where a view of the underlying rock could be obtained, he stated that the fossils were identical with those of the Murray cliffs.

Now, though there is every probability of the correctness of these conclusions, yet they cannot be considered much more than surmises without some further data, because Captain Sturt was not a very experienced geologist, and geology then was in no very advanced state, so that assertions on such matters must be accepted with caution. While stating this, a tribute is in other respects due to Sturt's merit; indeed, if it were not out of place here, I would add my mite to the general testimony of admiration for that learned explorer, whose zeal, untiring energy, and courage, were enhanced by a humanity and unselfishness rarely met with, and yet whose unpretending modesty shrank from praise, while it threw a charm over all his narratives. But still Captain Sturt may have been mistaken in the description of these fossils. It is not stated whether he made any collections of them; if so, they would set the question at rest at once. We will suppose, however, they were identical with those of the Murray. We may do so because of the nature of the country, and because of what is found to the south.

Following in this direction, we should come upon the Australian Bight, that is, supposing the deposits to continue from the parallel of Lake Torrens — a distance of about 5° of latitude to the most northerly portion of the Bight, and about 10° to the most southerly, which is King George's Sound. Now, be it observed, that to reach these points we should have to traverse a country with few or no elevations, and that as the rocks to the north were crag, and if those of the south were crag too, we may at least conclude that the strata are continuous,—at any rate, in some places. Of the nature of the deposit all along the coast we have the testimony of very many persons who have inspected it, and they all agree in describing it as cliffs of fossils, precisely similar to the Murray beds. In most cases, however, it is to be regretted, such accounts are from men not capable of examining whether the identity was a fact borne out by the fossils, or given at a time when geology was not so far advanced as it is at present; but the evidence will go far towards positive proof, when it is shown that the cliffs are precisely similar in appearance to those of the Murray, and struck the observers as being just such as would arise from a coral reef; for there is no other formation in Australia, as far as my knowledge goes, to which the same description would be applied.

Captain Flinders gave the following description of the Australian coast after his survey in 1802:—
'The length of these cliffs from their second commencement is 33 leagues, and that of the level bank from New Cape Paisley, where it was first seen from the sea, no less than 145 leagues. The height of this extraordinary bank is nearly the same

throughout, being nowhere less by estimation than 400 feet, nor anywhere more than 600 feet. In the first 20 leagues the rugged tops of some inland mountains were visible over it, but during the remainder of its long course the bank was the limit of the view.

'This equality of elevation for so great an extent, and the evidently calcareous nature of the bank, at least in the upper 200 feet, would bespeak to have been the *exterior line of some vast coral reef*, which is always more elevated than the interior parts, and commonly level with the high-water mark. From the gradual subsiding of the sea, or, perhaps, from some convulsion of nature, this bank may have attained its present height above the surface, and, however extraordinary such a change may appear, yet when it is recollected that branches of coral still exist upon Bald Head, at an elevation of 400 feet or more, this supposition assumes a degree of probability, and it would further seem that the subsidence of the waters has not been at a period very remote, since these fossil branches have yet neither been all beaten down, nor mouldered away by the wind and weather.

'If this supposition be well founded, it may, with the fact of *no* other *hill* or *object* having been perceived above the bank, in the greater part of its course, assist in forming some conjecture as to what may be within it, which cannot, as I judge in such a case, be other than flat sandy plains or water. The bank may even be a narrow barrier, between the interior and the exterior sea, and much do I regret the not having formed an idea of this probability at the time, for, notwithstanding the great difficulty and the risk, I should certainly have attempted a landing on some part of the coast, to ascertain a fact of so much importance.'*

It must be remarked that Cape Paisley, 124° E. longitude, is made the commencement of the Bight; but a reference to the map will show that the great bend commences farther east, nearly at King George's Sound. Captain Flinders also states in the above passage, that mountains were seen inland for the first twenty leagues, but this in no way affects the position as to the general continuity of the strata; for he further remarks, that the coral is seen in the cliffs at Bald Head, which is much farther to the eastward, and not very far from King George's Sound. The supposition of the receding of the sea, and the great reluctance displayed by the explorer to broach so daring a theory as the upheaval of the land, show that geological knowledge was not in such a state as to make observations in that department very satisfactory. The facts, however, are certain, and what Captain Flinders supposed to have been a barrier reef is the remains of something similar, though there is no water on the other side, as he and Captain Sturt at one time supposed, and the country is little more than a barren desert.

Captain Sturt, in speaking of the above opinion

* Flinders' *Voyage to Terra Australia*. London, 1814.

of Captain Flinders, says: 'His [Flinders'] impression, from what he observed while sailing along the coast, in a great measure corresponds with mine, when travelling inland. The only point we differ upon is as to the probable origin of the great sea-wall, which appeared to him to be of a calcareous formation, and therefore he concluded that it had been a coral reef, raised by some convulsion of nature. Had Captain Flinders been able to examine the rock formation of the Great Australian Bight, he would have found that it was, for the most part, an oolitic limestone, with many shells embedded in it, similar in substance and formation to the fossil beds of the Murray, but differing in colour.'*

Upon what data these latter statements are made, Captain Sturt does not tell us, but it may be here noticed, that he thought the beds were continuous with those he observed farther inland, that is, on the banks of Lake Torrens. Indeed, this is assumed, with good reason, throughout his whole work. It may be observed, that Captain Sturt differs with Captain Flinders as to the nature of the formation, and does not think that it arose from a coral reef; but the very facts he brings in support of his position bear out the conclusion he essays to disprove. From what has been already said in this chapter, the description of Flinders is just what we might expect from coral reefs, and had the exploration of the Pacific and the nature of our chalk rocks at home been better known than

* Sturt's *Central Australia*, vol. ii. London, 1849.

they were when Sturt wrote, he would not have contradicted the former explorer, as no one cared less than he did for his own opinion, provided truth were elicited. The peculiar views held by the latter were intended as a bold and ingenious explanation of the physical features of the interior, and are inconsistent with the idea of a coral reef on the coast. He supposed that the inland sea in the interior (of which there was probable evidence), gave rise to the crag beds, and what is seen on the coast is the drainage of this sea. I gather this view from his work generally, though it is hardly so distinctly enunciated; but, in speaking of the Murray and the granite rocks therein, he supposes that this on the Murray stopped the drainage, and gave rise to the deep deposit of exuviæ seen there.

There can be little doubt that these views are incorrect. The granite appears to me intrusive, and probably connected with the subsequent volcanic emanations which took place at Mount Gambier. They will be described hereafter.

To return to the evidence as to the nature of the Australian Bight. When Mr. Eyre went overland to King George's Sound, a Mr. Cannon was sent to Fowler's Bay (east longitude about 133°, south latitude about 33°), to meet the explorer with supplies, and survey the coast in the neighbourhood. He says:—' From the general flatness of the country, it may be presumed that *its characters* do not alter for a *great distance inland*. I observed nothing of the formation of the islands differing from the mainland, and I may mention that the rock of the Isles of St. Francis presented the same appearance as the Murray cliffs.'

I do not offer these and other similar testimonies as decisive of the nature of the formation, but they are rendered more than probable evidences when considered with other circumstances. These are:—The observations of Captain Sturt, the flat and open nature of the intervening country, so closely resembling the Murray Scrubs, and the absence of rivers which would lead to the inference of no rises or elevations, and, therefore, no upheaval or other likelihood of great changes, in the strata between the West and the great South Australian chain. It is not attempted here to define boundaries, or to say that there are no interruptions. How far the strata may extend to the north we know not; they are seen at Lake Torrens, but that may be its highest point, and its western extremity is unknown to me. As to interruptions, they are certain. The mountains alluded to by Flinders are instances; and then, again, the metamorphic rocks about Port Lincoln. Latterly, discoveries in the north-west country from Lake Torrens have shown that granite rocks and elevations occur in the interior. These may have been islands in the coral sea, or they may have been intruded subsequently, as is probably the case with the granite rocks in the bed of the river Murray, and through the Tatiara country to the north of Penola; but, at any rate, it is not pretended that no breaks occur in the strata, though there can be no doubt that, in spreading over as they do so wide an extent, the interruption is very small.

Having followed this formation westward of Adelaide, and having mentioned the evidence in favour of the surmise that a great portion of Central Australia is occupied by the same crag deposit, it is necessary to state how far it extends to the eastward of the boundary line between the colonies of Victoria and South Australia. It is rather singular that the 141st meridian of east longitude, chosen as a boundary (most unadvisedly, while the river Murray would have made a natural one) between the two colonies, should be really, within a few miles, a geological boundary. Generally, along the line, or near it, trap rocks occur, and continue for some distance. These rocks are merely a stratum, and are founded on the coralline rocks underneath. It appears that there has been, during the Post-Pleiocene epoch, an immense flow of basaltic trap, probably from submarine volcanoes, and this has given rise to the rocks as they are now found. Though there are a great number of extinct volcanoes to the eastward of South Australia, I do not think the trap has been derived from any of them, because the lava is less vesicular and more compact than can be accounted for by supposing a subaerial crater; and, secondly, there is no evidence of any great flow of lava from any that at present exist.

In some places, the limestone is seen to protrude

from the igneous rock, and at Portland Bay, about fifty miles to the east of South Australia, where the coast action has exposed a fine section, the coralline limestone is seen underneath. These beds I have carefully examined, and can state that the fossils are quite identical with those of Mount Gambier. There is certainly a greater proportion of one kind to the exclusion of others, such, for instance, as the abundance of *Spatangus Forbesii*, *Terebratula compta*, while univalves are almost absent, and the character of the stone is brittle and friable. But such differences as these, while the fauna remains the same, are more to be attributed to local circumstances than differences of geological position.

It would appear as if the beds at Portland were formed farther away from the reef than those at Mount Gambier, for corals are uncommon, the *Bryozoa* small, the fossils, with very rare exceptions, much broken, and the stone is more like the white mud, spoken of previously, than what is seen elsewhere. The *Cellepora gambierensis* is absent. The cliffs are topped with the oyster-shells seen on the Murray, and then overlaid again by the basaltic trap, which is here very much decomposed. The fossil cliffs extend along the coast between Port Fairy and Cape Otway, and this is about the same longitude as that on which they terminate on the Murray. Of their continuance in a southerly direction we have no direct evidence, but beds are described as occurring in Tasmania which bear a strong resemblance to our formation. In works on that country they are sometimes alluded to, but in Bunce's 'Australasiatic Sketches' they are more dwelt upon. As the passage is really very interesting, its quotation will be pardoned, as illustrating things which will be spoken of again, but I must admit that the phraseology of the description would be quite as applicable to primary as to tertiary rocks. In an overland journey to Launceston, speaking of a range of hills about twenty-one miles from Westbury, he says:—

'On ascending the ridge of this series of hills, a magnificent view presented itself, suddenly, to the delighted traveller, of rich and fertile surface, with purple-tinted romantic hills in the distance. After descending the ridge already named for the distance of five or six miles, we crossed the Moleside rivulet, so called from the circumstance of its occasionally disappearing and flowing underneath the ground, like the river Mole, in England. The whole of this neighbourhood is of limestone, with beautiful white veins, and the strata are nearly horizontal. There is a small circular plain, about the distance of five miles from where we crossed Moleside. The character of the country was most remarkable, and appeared intersected for many miles by numerous underground streams, flowing in different directions, and at various depths. The effect of these streams thus flowing underground causes the undermining of the superincumbent earth, which, being thus left without a foundation, has fallen in many places, forming pits and basins of the most singular kind, varying in depth of from twenty to two hundred feet, and shaped like a funnel. Many instances of this kind may be observed in the neighbourhood of Mount Gambier and many parts of the country near the Glenelg river in Victoria. In the bottom of most of them is a small circular pool of water, of immeasurable depth. A party on one occasion descended one of the deepest of them, and at the bottom found a cavern extending both ways, into which they entered. After following its course, a sound of running water was heard, and, although they were without lights, the reflection from the entrance was sufficient to enable them to distinguish a large body of water rushing from a height, and flowing away, as it were, beneath their feet.'

There is nothing here mentioned about fossils, and an apology is almost necessary for the introduction of a long extract which bears so little on the subject of this chapter. Still, the mention of the circular pits, the caves, and the underground rivers, is so very like what is subsequently to be described of Mount Gambier and the vicinity, that it was considered of some value to point out the resemblance, and thence the faint possibility of the deposit extending so far. If the beds were spread out to such an extent southwards, they would only be in the form of the white mud, which occasionally drifts to so great a distance from the reefs, and therefore fossils could not be expected. The mere existence of caves and rivers is, however, no evidence, as they are found wherever there is limestone. In the Wellington Valley, in New South Wales, there are many of great extent, containing bones, &c., though the formation in which they occur is very different from that of Mount Gambier. Some of these have never been explored, in consequence of the rush of wind from them, which prevents their entry with torches.

With the facts just mentioned all further clue to our formation eastward or to the south is lost, and though much uncertainty must prevail if we attempt to define boundaries, enough has been said to show that the formation covers a very wide extent of country. And now, having shown this immense extent to be covered by one formation, let us enquire what has been the nature of the operations which gave rise to it, and what other geological conclusions may be drawn from the detail given in the foregoing pages.

In the first place, the formation has arisen from a series of coral reefs, or is, in other words, the result of a coral sea. What has been already said in proof of this need not, of course, be here repeated; but, if there should be any doubt on the matter, what will be now adduced will serve to bear out the view already taken. Secondly, the land has been subsiding during the accumulation of the strata. There are many reasons to be given for this, but the most cogent of all is, that stratified beds of any thickness are never deposited except where subsidence is taking place. If we remark the coasts of Australia, which are now evidently in course of upheaval, only very thin beds are seen to result, because, as the raising force is constantly changing the coast or the shallow parts, where alone deposition takes place, there can be no time for any great accumulations. On the other hand, subsidence gives the greatest facilities for the deposition of thick strata, because, the deeper the sea, the more tranquil the bottom, and the greater area there is for the distribution of shells, either borne from the coast by currents or worn away from other rocks by denudation. Besides, the only favourable time for a developement of coral reefs is during a period of subsidence; at least, according to the ingenious theory now to be mentioned, and which is at present universally adopted.

Most persons are familiar with Darwin's clever and interesting work on the 'Structure and Formation of Coral Reefs,' but, as nearly all here adduced will be unintelligible without a clear knowledge of the subject, the repetition of the main points of the theory of that great geologist will be pardoned. In the Pacific Ocean and other tropical regions, coral islands occur of a most singular form: these are the atolls before described. It was formerly supposed that these were reefs built on the edges of extinct craters under the sea, and it was imagined that the ova of the coral animal pitched upon these sites as very favourable to their operations, and then commenced building a reef to the surface. There were circumstances apparently favourable to this view. An aperture was generally found on one side of the lagoon through which the lava of the supposed volcano underneath might have escaped, and earthquakes were not uncommon in this neighbourhood. But a fatal difficulty to be got over was their immense size; for craters were never known sixty miles wide. This was an enormous improbability. Besides which the depth around these lagoons was more than 1,200 feet, where bottom could be reached (which, was rarely), and it was known that the coral animal could not exist at a greater depth than 180 feet.

It of course suggested itself to every observer, that if there was an old crater underneath it must be at a great depth, and how was the animal to have got to it when a quarter that depth was sufficient to destroy its life? At last, Mr. Charles Darwin, after his voyage in the *Beagle*, suggested a theory which has been since universally adopted.

According to this, the coral animal seeks any foundation to build upon that is not out of its depth, supposing this to be an island. If the land be stationary, the coral will build to the surface, and then either die or extend itself in a lateral direction; if the land be in course of upheaval, the reef must at length perish; but if the land be subsiding, the animal will build to keep near the surface, and of course the rate of building must keep pace with the subsidence, or it will be submerged and destroyed.

It is evident that as the island sinks the distance between the reef and the shore will be increased, and when the land disappears entirely a ring of coral will still be at the surface round the remains of the old terra-firma, like a fence round the grave of the departed. It will be unnecessary to go through all the arguments by which this view is borne out, or to state the reasons of the breaks in the side, the lagoon, &c., but what bears on the matter in question will be elucidated as we proceed.

So that, from the theory just given, we must regard the coral district of the Pacific as an immense area of subsidence, with some few exceptions, and that the reefs there seen are memorials of high parts of the continent which formerly existed there. The reefs are of three kinds—barrier, fringing, and atolls (ring islands). Fringing reefs are those which surround continents and islands, lying close to the shore, with no signs of subsidence or upheaval. Barrier reefs are those which either extend along a coast line at some distance from it, such as the barrier reef of North-western Australia, 300 miles long, and sometimes 70 miles from the coast, or surround an island at a great distance from the shore, such as the reef which surrounds New Caledonia, so far from it that the mainland is invisible therefrom. The atolls have been already described.

Now it remains to enquire, to which class of reefs should the crag coralline formation be referred? It is obvious that very little of the original form can be traced from the rocks themselves, and therefore our conclusion must be drawn from analogy rather than from any other source; and thus it is to be inferred that what we witness is not so much the peculiar result of one or two coral reefs, but the remains of a coral sea of various reefs spread over a wide area, much as the coral sea of the Pacific is at present.* Now, if the bed of the Pacific were suddenly to be upheaved, so as to expose a sectional view of the rock which has been forming there during the existence of the present sea, what should we perceive? A very white limestone rock, fine-grained and soft, containing broken branches of the *Madrepora abrotanoïdes, Mœandrina dædalœa, Porites clavaria*, and other branched corals, while large *Pectens, Chamœ, Astrœa*, and various fishes' teeth, would be scattered through the mass amid *Coralline, Echini,* &c.

We should find deep layers of these, because the strata of the coral mud would be much more common, and cover a larger area than the reefs themselves, and these latter would have become nearly obliterated, as, in the course of time, their dimensions gradually contracted. The fossils we possess in the strata now noticed, and the conclusions they lead to, instead of indicating one particular reef, show us how the bottom of the Pacific would appear if now examined, and allowance made for differences of time, place, and species.

But how are we to account for the absence of

---

* It is to be remarked, that I use the word 'coral' here not in its strict sense.

the reefs themselves? This has for some time been a formidable objection to the reef theory of Darwin. It has been stated that no soundings, or only very deep ones, are obtained close to the atolls. Now, if the bottom of the sea were to be suddenly upheaved, immense coral mountains with basins on the top should result from the islands. But, wherever we have fossil evidences of an ancient coral sea, nothing like this can be observed; on the contrary, in the Australian coral reef, now under consideration, country wonderfully level is met with. To meet this difficulty, it must first be remembered that the state of the Pacific may be something peculiar to our era. This view is contrary to that which supposes all the phenomena of geology to have resulted from causes like those even now going on about us. But we may question the accuracy of applying the principle without some modification. We may ask, Do not the coal periods, the Wealden, the chalk, the crag, point to a peculiar modification of the earth, to certain kinds of vegetable and animal life, of which the remarkable growth of atolls is an instance in our own as distinct from other periods as the animals which now build them are distinct from similar genera in former epochs?

But, even waiving this explanation, it must be remembered that the bottom of a coral sea is never thus suddenly upheaved. The deposit, after slowly subsiding, may remain stationary long before upheaval takes place at all. It would then be submerged, and subject to the action of the ocean. This would stratify and wear away any eminence, to say nothing of the slow rate of upheaval which would give ample time for aqueous erosion or denudation. It has been remarked, that the atolls are gradually growing smaller as the bed upon which they rest is subsiding. Now, if this be the case, a time must come when they are reduced to a mere peak, and then the animals will cease building. The subsidence must be very slow for the polypi to be able to keep pace with it, and so once the coral was dead the atoll would be for an immense time exposed to the fresh ravages of the ocean, which would not be long reducing it to the level of the rest of the sea bottom. The very form, indeed, of atolls shows them to be liable to rapid destruction, once the building operation of the Zoophytes had ceased. Soundings taken close to them show that they descend in the form of a very steep narrow cone; and to find such in existence after long submersion beneath the sea and subsequent slow upheaval, would be quite unaccountable, being so inconsistent with our present notions of the sea's ravages, and we should have to frame some hypothesis why the ocean, which in some places tears down lighthouses, rends rocks, and destroys massive and gigantic breakwaters, should spare such a fragile structure as a cone of *dead* coral. For, as long as the polype lives, it can build up faster than the waves can destroy; but take away this force of animal life from the struggle, and it is not difficult to see with whom the victory would rest in the end. Yet, in spite of these observations being true, we might meet with such remains. The hypothesis might be in the main correct, with occasional exceptions, though what follows decides little either way.

Where evidence is very weak it is almost useless to adduce it; but it may be of some service to describe two basins of limestone in the interior, which it is just barely possible may be relics of atolls. They both occur in the Tatiara, at about 120 miles north by west of Mount Gambier. One is called the Swede's Flat, and will be described at length when we come to treat of caves. It is a flat plain, like a dried-up lake, as level as a bowling-green, fourteen miles long by two broad, and completely encircled by hills. There is no considerable declivity outside it, but the soil is sandy and composed of coralline rock. If it had been a lake, there would be some traces of streams by which it became filled, but there are none. It never retains water on its surface, in consequence of large holes which drain underground.*

The other is a hill about three miles slightly to the west of the southern end of the one just described: it is on the road between Kelly's and Lawson's stations, and is called the Half-way Gully. It is like an immense crater, with a break on one (the eastern) side, as if for the passage of lava. It

* The level of the bottom of this flat is much *above* the surrounding country.

is densely covered with brushwood, through which the coralline limestone peeps from time to time. Standing on its edge, the depression of the centre appears about half a mile wide, and no other rock but the coralline is anywhere visible. There is no sudden declivity from the side, but the same scrub and sand. Probably the hill is over a thousand feet high, but it is joined to the range, and, therefore, very little elevated above its neighbours. There are only two things to which it could be compared — a crater, or an ancient atoll; and, as there are no trap rocks within miles, we might, though perhaps on weak evidence, suppose it to have been the latter.

In the next chapter will be described some of the vicissitudes which happened to our reefs, after their burial in the bed of the ocean, which indicate that its upheaval was neither very rapid nor very soon after the submersion of the reef; what has been said will, however, be sufficient to prove that we may regard the Australian formation as the result of a former coral sea, without expecting to find the remains of either barriers, reefs, or atolls. Let it be further remarked, that if it would apparently require an immense layer of time to destroy all trace of the structures to which the beds owed their origin, at least, it would be far less than is ordinarily required to account for geological operations; and that while thousands of years would go as nothing to explain the coal formation, or the upheaval of the post-glacial beds, a very few thousand years would be sufficient to reduce the strata to the state in which they are now seen. It has been already mentioned, that a thick stratum of limestone, without fossils, is generally the uppermost bed, and this may have been derived from the wearing down of the reef, and the rearrangement of the upper part of the beds. The absence of this deposit in certain places does not affect the general argument, that its absence was rather the exception than the rule.

I should imagine that the reefs must have existed more to the northern part of the present formation. This might be supposed from the greater warmth of the climate, which would be more favourable to such growths. Besides, the fact is apparent in the strata. The farther south they are traced, fossils become less frequent, more broken, and the corals are much fewer. At Portland Cliffs, as I have already said, corals are very rare, shells uncommon, and much broken, while the limestone is more like the dried white mud already frequently alluded to. At Mount Gambier (distant seventy-three miles), corals are more common and less broken, shells of frequent occurrence, while sharks' teeth (*Oxyrrhinus Woodsii*) are abundantly distributed. At Mosquito Plains (seventy miles from Mount Gambier to the place referred to here), the corals are the predominating fossils, and are very perfect, preserving all their beautiful manifestations with little or no breakage, and often upright in the position in which they grew. At the Murray, the fossils are principally shells of large univalves and bivalves (the former principally

Turritellidæ) showing a much more tropical fauna, though the coral is not uncommon, with a greater variety of species. I do not know of any place where the variations of temperature can be traced better than in this district, and the approach to warmer latitudes is as clearly marked in the fossils as the approach to the snow line or a mountain is marked by the flora.

It is not often we have an immense formation left undisturbed, so that we can journey from one end to the other, and speculate on the exact temperature which prevailed, by a comparison with the present habitat of similar species. Whether there is any great variation from what at present obtains, can only be gathered from a more minute examination of the faunæ than I have been able to afford them. That, however, the mean temperature of the sea has been greater than what it now is near the same places, there can be no doubt; but this does not necessarily imply that the mean temperature of the earth was greater. Everyone is aware how much the warmth of a climate or a sea is dependent on the distribution of land in the vicinity. Thus, the current of water from a tropical latitude may be turned aside by a peculiar conformation of the coast, and so carry a sea of almost tropical temperature into a temperate zone. Such really happens in the Bermudas (the only extra-tropical locality where coral reefs flourish), where the Gulf Stream keeps up a very high temperature; and, as in this part of Australia, the land may have been so disposed near what was then the coral sea, that a very high temperature was kept up.

It is perhaps useless to speculate where the land was at this time, but we may at least offer some suggestions which bear slightly on the question.

A large area of the bed of the Pacific is known to be subsiding, while, on the other hand, an immense portion of the continent of Australia is known to be uprising. If what we know to have been sea during the Crag and probably Post-Pleiocene period was subsiding, was the present subsiding area of the Pacific a continent? What would be its position if it were so? It has been observed that it is mostly within the tropics. Now, Sir Charles Lyell has beautifully shown in his 'Principles of Geology' that a distribution of land exclusively within the tropics would have given rise to a temperature perhaps beyond human endurance. Even supposing that the land was not exclusively within these limits, a large tract of land, equal to the present coral sea, would have materially affected the temperature of our extinct sea, and amply account for the existence of our almost tropical shells in temperate latitudes. It is true that even in the coral sea there are occasional periods of upheaval at present, but these are very small, and do not affect the general position that a large tract is subsiding.

It would be a very curious phenomenon, if it were proved that the upheaval of one part of the earth's surface was compensated by the subsidence of another. It would show a regularity in such movements that might eventually prove them to result according to a fixed law. I have always been of opinion that disturbance of the kind was less common and more general in the Southern than in the Northern hemisphere, but this is little more than opinion. At the same time, the multitude of different strata in Europe, the small area they occupy, and the marks of upheaval, contortion, and breaks at almost every step, while large undisturbed tracts exist in South America and Australia, would seem to confirm the opinion.

I cannot close this chapter without alluding to a train of thought into which I have been led while studying these rocks.

The three great periods into which fossiliferous rocks have been divided are characterised by distinct predominance of one kind of animal life. Thus, the Palæozoic rocks have been characterised as the age of sauroids, fishes, and articulata; the Secondary—reptiles, marsupials, and cephalopodous mollusca; the Tertiary—pachydermata, and by a gradual approach to what the earth is now. It is not pretended that no other living things existed, but that these predominated, and, though great additions may hereafter be made to our list, by the discovery of new animals, it will not affect the general proposition, that a peculiar class of animal life was more common in one period than another. It has been further observed, by many geologists, but more especially by the late Hugh Miller, that a gradation in creation may be traced, in the three enumerated periods, where a series of less perfect organisations seem gradually to prepare for higher creatures.

It should here be remarked, that when the term 'less perfect' is used, it is not meant to be accepted literally. All God's works are equally perfect, and there is as much room for wonder and admiration at the perfection of design in the simplest plants as there is in the most complicated animal; but what is meant is, that some organisations are less complex, or have less special adaptations than others. Thus, the simple *Acalephæ*, or jelly-fish, perform the functions of respiration, absorption, assimilation, and circulation, in the mass generally, and each function is performed perfectly, to meet the requirements of the animal. In a warm-blooded animal these functions have all special organs, such as the lungs, the lymphatics, the stomach, the heart. These are more complex, but not more perfect, or fit the animal better to fulfill the end of its creation.

But there has been (so geologists tell us, admitting their knowledge to be very imperfect and unsatisfactory as yet) a developement from the earlier periods of an approach to a more complex organisation, which ended in Man. Thus, in the sauroid period, the kind of animal most common was one with a very low cerebral developement, an imperfect respiratory system, a heart with only one auricle and ventricle, and whose ova were extruded in the most elementary state, germinating and developing themselves quite distinct from the parent. Birds were also common in this period, and these, though possessed of a complete respiratory and circulating system, are of a very low cerebral developement, and the young are born in a most embryonic state. The Articulata common to this epoch are low in the scale of animal organisation. Fishes give way to Reptiles, whose respiration is by lungs, whose heart contains three cavities, whose ova are more developed than those of fishes (some are viviparous, *e.g.* land salamander). Articulata give way to Cephalopoda, the most highly organised of the Mollusca. Lastly, the first Mammalia appear, but these of a very low organisation; for the only ones that have been found have been proved to be marsupial, and these are animals only one degree removed from birds. Their cerebral developement is low (the corpus callosum being most rudimentary, and the convolutions of the brain entirely absent), while their young are developed distinct from the mother, being born in a merely embryonic state. Finally, in the tertiary period we see the highest developement which the animal world has undergone up to the time of man's creation. The commencement is seen in the earlier strata becoming gradually more numerous up to our own period.

Now, it has been remarked by some geologists, that, from the fossil flora and fauna of Europe, during the Pleiocene period, a state of things must have prevailed there very similar to what obtains in America now. So that America is, in reality, in her Pleiocene period, or one geological period behind the Old World. This principle is of course meant to be accepted in a modified way, for there must be a great admixture in the living animals and plants between it and those which now tenant Europe. But, as a general principle, it has been stated as possessing some truth, and several remarkable facts in accordance have been given.

Now, with regard to Australia, I wish to enunciate a similar principle (not orginated entirely by me), of course subject to many limitations, though less than what are required for America, and decidedly more marked in character. I believe that the present state of this part of Australia is very similar to what Europe was immediately after the secondary period, and that really, in regard to the developement of its fauna and flora, this continent is far behind the rest of the world. The position of Australia renders it less liable to an admixture of its species with those of other continents, and, therefore, its natural history is, to a certain extent, peculiar to itself.

In the flora, the correspondence to the secondary period is well marked. There the *Araucariæ*, so common to the secondary rocks, are represented, and these are only found in Norfolk Island and Australia. There are the *Zamiæ* and *Arthrozamiæ*, found only at the Cape of Good Hope and Australia, being closely allied to *species* found in secondary deposits. There are likewise plants which, though not connected botanically with the mesozoic flora, bear a striking resemblance to them, and these are the *Xanthorrhœæ*, which abound in all the continent. If we may judge from the few specimens which have been preserved to us from secondary rocks, the flora was not abundant there, and in this particular our country resembles it; for the general character of the country is most decidedly barren, more especially in those places where the strata I have been describing are found.

Then, with regard to the Mammalia, no indigenous animals have been found distinct from the Marsupialia, except rodents, and one or two species about whose introduction doubts have been entertained. The rodents are an order which has many affinities with marsupials, and one species occurs where the characters are interchanged,—the Phascolomys. Our birds, though beautiful, are comparatively few in number, and even these not all peculiar to our country. And, lastly, there are no secondary rocks found in Australia,* but a great portion of the country appears to have been recently raised from the sea, where it has undergone a state of things very similar to our chalk; while the immense tracts of desert country, and the large portions that are quite unavailable, indicate a territory less adapted

---

* In the 'Geological Society's Journal' for May 1860, there is a letter from Mr. Selwyn to Sir R. Murchison. In this it is stated that two fossils have been found, which Professor M'Coy thinks to be decidedly belonging to the chalk.

for the habitation of man than any tract of land of similar size on the face of the earth.

The following passage from Mantell's 'Wonders of Geology' is directly confirmatory of the previous statement. Speaking of the Wealden strata, he says:—' Nor can we resist the conviction, that not only did the same terrestrial area, however modified it must have been during the long succession of ages, supply the débris of an almost unchanged system of animal and vegetable life to the jurassic seas at first, and subsequently to the cretaceous ocean, but that also the fauna and flora of this ancient land, of the secondary epoch, had many *important features which now characterise Australia.* The Stonesfield marsupials and the Purbeck plagiaulax are allied to genera now restricted to New South Wales and Tasmania, and it is a most interesting fact, as Professor Phillips was first to remark, that the *organic remains* with which these relics are associated *also correspond* with the *existing forms of the Australian continent* and neighbouring seas; for it is in those distant latitudes that the waters are inhabited by *Cestracions, Trigoniæ,* and *Terebratulæ,* and that the dry land is clothed with araucariæ, tree ferns, and cycadeous plants.'

We have no means of knowing what was the state of the earth after the secondary period in Europe; but, knowing the conformation of the underlying rock, and knowing that it is of a kind that must give rise to a sandy soil, taking, perhaps, ages before a good loam could be formed upon it, and knowing, further, its generally level character, there must have been a state of things after its upheaval not very different from what is now observed in Australia. The absence of secondary rocks in the continent, as far as it has been yet explored, is certainly remarkable; and though negative evidence must not be relied on, still sufficient of the colonies is known to make their absence a matter of some certainty. Abundance of the rocks formed, *i.e.*, metamorphic slates, schists, eurites, porphyries, and granites, are found principally on elevated tracts or mountain ridges. Trap rocks are common, mostly in Victoria. Silurian slates have been found, I believe, in all the colonies, as well as the old red sandstone coal measures (excepting in South Australia), and in some places the new red. But after these there is a great hiatus, and the tertiary rocks are the only continuation that we have.

Now, I do not mean to say, when I state that Australia has gone through a period corresponding to the Secondary in Europe, that there is a blank in the geological history of the continent, or that our primary corresponds to the secondary elsewhere. There may be, for all we know, mesozoic rocks underneath our cainozoic, though it is not probable, or Australia may have been a group of islands or a continent during all that time. Perhaps it would be better, to prevent misapprehension, and to avoid the infinite limitation which a more general principle would require, by stating, that no more is meant in the above facts than this,—that the state of things now seen in Australia bears a strong analogy to what Europe must have been at the close of the mesozoic epoch, and that, in accordance with this, Australia, in its natural history, is far behind in developement to any other part of the known earth.

If any further speculation on the subject should cause the principle to be more strictly applied, the cause of science will be more benefited that way than by stating too much in the first instance. Not to conceal, however, any fact which might militate against my view, it must be mentioned that the fossil remains of a lion and of a native dog (supposed at one time to have been introduced) prove that until recently these animals flourished as indigenous to the soil.

If the above principle be true, it gives rise to one important consideration. There are some who, when they discover a sequence in creation, do not trouble themselves to enquire why the Almighty chose such and such a plan in following it out; they look upon each part as of necessity belonging to the whole, and they would seem to infer that, because God adopted a certain gradation, no deviation was possible. Something of this kind is to be traced in an otherwise able work, entitled 'Vestiges of Creation,' and many other works on creation imply the doctrine, that if the motive of the earth's creation were to prepare a habitation for man, all that went before manifestly tended to this, and at no other geological period than at present could he have existed. Undoubtedly the present period is better for man than any which went before, as far as we know, but it is far from being impossible for him to have existed previously. The very fact that man finds an easy, nay, comfortable, subsistence in Australia, which, whether my principle be admitted or not, is far behind other countries in natural developement, proves, on the one hand, the perfect adaptability of the earth as a residence for man at other periods — besides our small conception of the plans of the Creator; while, on the other, the better adaptation of the other parts of the earth, more advanced and developed, proves the beneficence of the Author of it all in perfecting man's habitation to the highest degree before He placed him upon it. We may, however, say, that speculations which threaten large conclusions are best not made when they militate against certain truths (*i.e.* revealed) on such weak evidence as the *negative* evidence of geology. A limited amount of reflection will show how very small a portion of the earth's products get buried in marine or fresh-water strata; and yet how small, how very small, a portion of even these do we know,—how little, out of such a mass, can a few quarries, a few wells, a few mines or excavations, tell us! What, then, is the value of such theories as those which rest on the *absence* of our knowledge for their principal argument?

When we contemplate the vicissitudes to which the earth has been subject, we cannot help tracing a progress and civilisation in nature similar to what men pretend to in history. An ungenial soil has arisen from a recently-raised tract of land, and this, again, only supports a most meagre amount of animal life, not only few in numbers, but low in the scale of vital organisation. What a difference would have resulted from a different fauna is told by the experience of those who have kept sheep and cattle stations, who say that every year after their introduction sees an improvement. As it was, however, prior to the entrance of the white race, while nature toiled towards a better state of things, the land was ill adapted as a residence for man. Human nature actually languished in many parts of it, that is, were driven to means of subsistence badly adapted to support life. Captain Sturt mentions that the natives of the interior depended almost entirely on the grass-seed for their support, and even these sometimes failed, so that they were met with in a most emaciated condition. The same author also mentions, in his account of the journey to the Darling in 1828, which was a season of great drought, that most of the natives were nearly starved, a whole tribe having to subsist on the gum they collected from the bark of the wattle-tree. It would appear that Australia is subject to periodically dry seasons, and that then the natives of the interior suffer fearful privations, and become subject to contagious disorders. But this does not

L

interfere with the fact, that the continent *can* support human life, for, in spite of all their endurance, they did not appear to be diminishing in numbers until the white race came among them.

It is true that distress and famine only occur among the natives of the desert part of the country, while those in the better lands near the coast and mountains lead a more comfortable life. But the state of the soil and the country generally has a great deal to do with the fact, that the aborigines of Australia are, with the exception of the Fuegians, the most degraded, the most helpless race on the face of the earth.

In conclusion, let it be added, that if this country ever becomes great among nations, it will not be owing to the possession of many natural advantages. It is melancholy to look upon the map, and think of the immense tract of soil that must ever be useless to man. A bright future may be in store for some places, it has already dawned upon others, but to think of the vast deserts, sometimes bordering close on the comparatively small tract of agricultural land, leaves but little hopefulness for the greater part of the continent. There is room, however, for years to come, for settlers, on spots as rich, perhaps, as any the earth affords. But from the interior we turn in a despair like that of Captain Sturt, who, when he had penetrated to the farthest point ever reached by the European, stood upon a mound of sand gazing, as he said, upon an expanse unequalled in the world for barrenness and desolation.

We gain, from time to time, some trifling increase to our knowledge of its aridity, but all our knowledge results in this, that we know it never will be a home for man—that all our efforts at its exploration have been baffled, and the explorers sent back sickened and exhausted—that the bones of one of a finely equipped party, led on by an indefatigable genius,* now whiten on some part of its arid expanse.

* Leichardt.

NOTE TO CHAPTER V.

Since the above was written, Mr. Stuart has succeeded in reaching the centre of the continent, and had nearly crossed, when he was driven back by the natives. He has started again with a party of thirteen men. The loss of Burke and Wills will be also fresh in the memory of all.

I have lately heard that the bones of an extinct hyæna have been found in a cave here. This is very doubtful. Large bones were also found in some clay which fell away from the banks round Lake Colac. I could never ascertain what they were, nor what ultimately became of them. If the banks round these lakes owed their origin to a process similar to the banks round lakes described in the last chapter, they must have been land animals or amphibious reptiles to have become imbedded in them. I am assured, however, by those who saw them, that they were much too large for any kangaroo. Could it be the *Bothriceps australis*, described by Professor Huxley in the *Geological Society's Journal*? He states that the specimen was in the British Museum, but that nothing more was known of it than that it came from Australia. The animal is a reptile, but probably not geologically modern, and more likely came from a sandstone formation.

CHAPTER VI.

HOW THE REEF ENDED.

CESSATION OF THE CORALLINE FORMATION.—DESCRIPTION OF UPPER CRAG.—EXTENT OF IT.—DERIVED FROM AN OCEAN CURRENT.—GUICHEN BAY BEDS.—ABSENCE OF FOSSILS IN THEM.—CAPE GRANT BEDS.—STRATA THERE DESCRIBED.—TRAP ROCK AND AMYGDALA.—SIMILARITY OF UPPER BEDS TO UPPER CRAG IN ENGLAND.—SINGULAR FORMATION NEAR THE TRAP.—LOCALITIES WHERE THE UPPER CRAG IS FOUND.—BROKEN FAUNA.—REEFS LEFT OF CRAG.—CONCRETIONS.—NOT OWING TO CASTS OF TREES.—DECOMPOSITION OF THE ROCK.—BLOWHOLES.—DENUDATION AND UPHEAVAL.—WHAT BECOMES OF DETRITUS.—HISTORY OF THE DEPOSIT.—DENUDATION.—CORALLINE CRAG OF SUFFOLK.—WATER-LEVEL.—DEEP-SEA SOUNDINGS.

FROM the consideration of the coral beds we pass to those next in succession above. Fortunately, there is no blank in the geological history of the country now treated of. What we next find is the sequel to what has gone before, being just, in fact, what we might expect would, in the operations of nature, succeed the formation just described. It has been seen, that, during the building of the crag, the bed of the sea was subsiding, and that before the subsidence was terminated, or at least before upheaval commenced, the coral animal must have ceased building. Many things might have caused the destruction of the zoophytes. Either a sudden subsidence, or a change in the temperature of the waters, or the advent of a current containing sediment. That this latter cause is fatal to the progress of reefs is proved by observation. One of the causes why one side of the atolls is invariably broken down (a circumstance appealed to by earlier geologists in favour of the view that they were extinct craters) is because of a current of sediment in that direction being caused by the prevailing wind. In fringing reefs the parts opposite a river or stream from the land are nearly destitute of coral, and what is near them is always dead. Breaks in barrier reefs are traced opposite streams in the land which they surround, even long after they are so far removed as to cease to be affected by them. All or either of the causes above enumerated may have combined to destroy the reefs here treated of, and probably the latter bore the chief part. It is of no moment to enquire now how it ceased to exist; of one thing we are certain, and that is, that the deposit did terminate and a change came. This change gave rise to a different kind of rock, and this is what next comes for our consideration, and forms the subject of the present chapter.

Round the coast (which, as before stated, principally consists of coralline cliffs or hillocks) patches of a different kind of rock from the white chalky deposit are occasionally seen. At times, it forms sea-cliffs of itself, and then it affords a good variety, from the generally uniform white coast line, this being dark brown in colour, and more compact and rugged than the underlying strata. It is found more or less all round the coast of the colony of South Australia, and perhaps it extends all along parts of the Australian Bight. It is seen to most advantage where the coast is bold, and where it forms cliffs; and, as a better idea can be gained of the nature of the formation from such localities, I shall confine myself to them for the present. The principal places, then, where the rock is observed to most advantage are Guichen Bay, a port on the South Australian coast, between the most southerly part of the colony and the Coorong, and an indentation on the coast between Cape Bridgewater and Cape Grant, a little to the west of Portland, in the colony of Victoria. The whole eastern and northern sides of Guichen Bay are composed of low sand-hills, scarcely thirty feet above the water-level, but on the southern side a change takes place. The sand is replaced by rough craggy rocks, which, though not rising very high, are bold and abrupt, sometimes presenting a perpendicular face to the heavy surf which beats upon that coast. Seen at a distance, one would imagine that these rocks were divided only into larger strata, fourteen or sixteen feet thick, but, on a closer inspection, another kind of stratification is discernible. In addition to the great divisions, which are so distinct that one could almost suppose that they were huge slabs of rock laid upon one another, there is cross stratification. This is a lamination which divides the beds into strata about two inches thick, but they are never horizontal like the real strata, are scarcely ever parallel, and never continuous across the great divisions which divide one bed from another.

Now, all these appearances taken in connection with the mineral character now to be described, are clearly indicative of an ocean current. Anyone conversant with the elements of geology will not require to be told why this conclusion follows, from the facts above stated. It will be sufficient to say, that the want of horizontality in the smaller strata is due to some disturbance in the water from which they were deposited, and, as they bear in one particular direction, this must have been owing to a stream which deposited particles as it flowed along. The greater divisions are caused by an alteration in the direction of the current, which, before it would deposit any new matter, would carry away the lighter superficial particles, and wear down to a smooth surface all the inequalities left by the former stream.

If there were any doubt about this theory, it is quite removed by the nature of the rock. We know what kind of matter we should expect to find at the bottom of such a stream. Their course is generally slow, and therefore only small fragments of shells, grains of sand, and fine mud would be carried by them. A river will carry down mud from the banks, and fragments of wood, but an ocean current, which generally takes its rise in deep water, can only have the detritus of the rocks and shells it has acted upon.

The material of the rock now under consideration would appear, at first, to be a coarse-grained sandstone. Under the microscope, however, it is found to consist of small particles of shells, worn by attrition into thin scales, and small grains of a quartzose sand. It is freely acted upon by acids, and, with the necessary reagents, shows great quantities of lime, magnesia, and silica, with traces of sesquioxide of iron and sulphate of lime, but no phosphates nor organic matter. There are no fossils, excepting in a few places which I shall specify. From these facts, therefore, we may not only conclude that the deposit was from an ocean current, but also that it was a considerable distance from any land; because, coast drifts are generally rather rapid, being derived from large rivers or similar causes, while those far from land seldom exceed the rate of three miles an hour, and anything much quicker than this must infallibly have included larger particles of shells, and even whole ones.

Guichen Bay is not so well provided with this rock as a small inlet at the south side of Cape Lannes, the promontory which helps to form the south-east side of the harbour. Here the rocks are seen in bold section, for sometimes the cliffs are nearly a hundred feet high.

The little bay is very deep, so that the water washes the cliffs nearly all round. In some places the action of the surf has undermined them, and caused them to fall, and the spray has eaten into its soft friable texture, giving parts a wild and jagged outline. These features, and the singular cross stratification of the cliffs that have escaped the ravages of the ocean, the dark hue of the stone, the heaps of ruins scattered about like fallen castles, and the boiling of the heavy surf, which, even in the calmest day, breaks upon the rocks, make a sublime scene, which for wild beauty would be unequalled in Australia, were it on a little larger scale. Even as it is, however, it reminds one of the bold coasts of the Highlands; and the little verdure which the mesembryanthemums give, as they creep down the surface of the rock, or hang swaying on the wind, tends little to soften its desolate and savage aspect.

There are, as I have stated, no fossils; but the summit of each cliff is topped by a stratum of compact limestone, horizontally disposed, but lying unconformably. This, I presume, is a relic of the last coast action, before the deposits were upheaved to their present position; and from the fact that the same stone, lying in the same manner farther inland, contains marine fossils of existing species, I have no doubt that it is of the same age as the very recent beds to be spoken of hereafter, which exist all round the coast.

Let us turn now to the other locality spoken of above, as possessing the same beds. This is the bay (Grant Bay) between Cape Bridgewater and Cape Grant, a little to the west of Portland Bay. There the deposits are seen to greater advantage, and on a larger scale; besides, from the rocks associated with them, we are able to decide more positively as to the total distinctness of the formation from the coral reef. The bay is some three or four miles wide, bounded on the two sides by the capes above mentioned, which are fine rocky headlands, with deep water at these very places. The whole coast of the inlet is very precipitous,—so much so, that there are only one or two places where you can descend from the cliffs on to the beach. Seen from above, the appearance of the bay is peculiar; because, after a little belt of sand, all round between the cliffs and the water, the surf beats in amidst a confused mass of large black boulders, much corroded by the action of the sea, but still preserving an irregular figure. These are trap rocks, or a very vesicular brownish-black basalt. To this the precipitous coast forms a great contrast, being the yellowish-brown stone, precisely similar to what is found in Guichen Bay, but it is seen to be based on the same basaltic rock throughout its continuation.

But the stratification is what makes it most singular. Not only are the minor laminations quite out of the horizontal, but even the great divisions. One part of the bay looks as if the strata had been deposited horizontally, and that afterwards the two ends had been upheaved and pressed close together, making the beds almost like the letter W, only a little more rounded. Most frequently, however, the great divisions preserve a serpentine line, though, in places, the diagonal or cross stratification is alone irregular, the rest preserving a uniform horizontal line.

Descending to the beach, the brittle and friable nature of the stone is observed, it being less compact than what is seen at Guichen Bay, and this is what makes the descent to the coast so difficult, for the mere action of the weather has eaten away the face of the cliff, making the summit in many places overhang the base by many feet. The rock is of the same texture as that of Guichen Bay, a calcareous sandstone, in every respect similar in appearance, and, like it, containing no fossils. At first sight, one would be induced to refer the contorted appearance of the beds to irregular upheaval, which has twisted them and bent them into the inclined position they hold at present. But this theory is quite untenable. Whatever upheaval there has been was of a most regular kind, and equal in its operation.

The underlying strata are in no way disturbed, and the stratum of basalt upon which the sandstone (or upper crag, as I shall term it,) rests, is as horizontal as the sea. The only way, then, of explaining the irregularity, is by supposing the current, from which the detritus was deposited, to have been rather strong and variable at this place, giving rise to shifting sandbanks, more abrupt in form, and more liable to change their character, at every change in the direction of the stream.

Perhaps, again, the sea was shallower here; and

just as there are sandbanks along the coast, with deep soundings round them, liable to change their form every year, so these beds may have existed near a coast and been subject to great vicissitudes. The layer of trap upon which they are based affords a good answer to a difficulty I at one time felt about the upper crag, as I shall term it in future. It is well known that the summit of dead coral reefs presents an appearance very similar to the detritus borne down by an ocean current. Thus, in Henderson's Island, described by Captain Beechy, mention is made of its being an upraised coral island, and is thus described by Lyell:—' It has a flat surface, and on all sides, except the rock, is bounded by perpendicular cliffs, about fifty feet high, composed entirely of dead coral, more or less porous, honeycombed at the surface, and hardening into a compact calcareous mass, which possesses the fracture of secondary limestone,'* &c.

The cliffs are considerably undermined by the action of the waves, and some of them appear on the eve of precipitating their superincumbent weight into the sea. Now, though this description differs from the deposit under consideration, inasmuch as it speaks of dead coral interspersed through the mass, yet the general character of the rock was so similar that there was some possibility of its owing its origin to an upheaval of the coral rock as a dead mass. Those at the rim of the lagoon in Atoll Islands are described as being a

---

* *Manual of Geology.* Fifth edition.

mass of calcareous sand, heaped together with broken shells and other débris. Of course the texture of the crag was rather against such a supposition, but then it had been lying immediately above the coral rock, and sometimes containing larger portions of shells, with an occasional fragment of what appeared to be coral, only too much broken to be certainly classed as such. The only thing that could settle the question was the existence of some intervening rock, which would show them then to be quite distinct from one another. This was found in the stratum of trap rock, upon which, as I have already said, the cliffs of crag rest.

At one time, I could hardly imagine that the trap really was underneath, because at Portland Bay, a short distance off, basaltic rocks are seen on the top of the highest cliffs, and, though these are of coralline rock, yet, as there are extinct volcanoes (Mount Napier and Mount Rouse) in the neighbourhood, I imagined that the trap rock was lava, which had flowed over the cliffs at Portland into the sea, at a time when the coast had assumed its present figure, and that the accumulation at Cape Bridgewater was what had flowed to the bottom of the cliffs, and been stopped further progress by them.

And it was very difficult to arrive at the real conclusion, because, on descending into the bay now described, the basalt could be traced to the foot of the other rock, and then seemed discontinued, and the ordinary limestone took its place.

But a minute examination showed that the break in the strata was more apparent than real. What seemed to be the crag was, in fact, a mere coating of limestone, which was washed down by the rain and by filtration, so as to completely cover the trap rock underneath, and make it appear like limestone. A very little digging, however, completely removed it, so as to show the vesicular volcanic rock beneath. It was curious to remark how the lime had been washed down, so as to form a sheet of stone over the underlying strata. This, too, was done much in the same way that stalagmite is deposited at the bottom of caves; and when we bear in mind that every particle of the stone must have been dissolved by water and then redeposited, no moderate amount of time has been consumed in forming the large crust of stone found in the bay now described.

The trap is quite amygdaloidal, that is, every one of the vesicles in the stone has been filled with crystallised carbonate of lime, rounded in the form of the mould in which they occur, and of a translucent yellow colour. Some are of a reddish tint, from the presence of iron, and they generally radiate out in beautiful acicular crystals, from the centre to the circumference; but more commonly they are small, and, wherever a piece of the rock is broken off, it appears as if studded all over with minute wax lentils. It is quite extraordinary how the lime filters through the stone to form crystals in the crevices. Break off the rock where you will, and, how compact soever it may be, the centre is sure to be impregnated with lime.

At the foot of the trap a very singular formation is seen. The sesquioxide of iron, washed from the volcanic rock, has acted as a cement to large particles of shells, rounded, water-worn, and of large size. This has formed a conglomerate of intense tenacity. The fragments which protrude look as if they could be picked off, but a great exertion of strength will not detach them. The shells preserve their colour in most instances, and the conglomerate is like a very pretty mass of flower petals. Mr. Darwin mentions a similar deposit as occurring at Ascension Island.

To return to the upper crag, there can be little doubt, then, that it is quite distinct from the coral formation, and that it is identical with the deposit found at Guichen Bay, because it lies like it directly over the crag beds and under a hill, hereafter to be shown as more modern deposit occurring in both places.

There are other localities in which the same strata are found, as, for instance, on the coast at the mouth of the Glenelg, in Victoria; again at Rivoli Bay, south of Guichen Bay; at Lacepede Bay, to the north of the same place; besides on many reefs of rocks that rise out from the coast, and in spots here and there, scattered more or less all over the district. Some of these latter possess peculiarities worthy of note. They are generally on rising ground, lying immediately above the coralline strata. This position clearly shows that other parts there have been which denudation has removed, and the texture of the rock itself, in such cases, gives the reason why it was able to resist ravages which destroyed the continuance of the beds. Thus, at Mount Gambier, at the edge of a deep circular chasm, where the upper part of the rock has fallen into a hollow, caused by the erosion of underground streams, there is a good section of the beds exposed. On the top of all are seen the ashes from the extinct volcano in the vicinity; underneath is this deposit, about sixteen feet in thickness. Its appearance at a distance would lead one to imagine that it was full of shells; but it is not. There is nothing but a mass of broken testacea, so confused and so broken that I have never been able to recognise one of them, with the exception of a large *Ostræa*. The face of the stone is perforated on all sides with the borings of the *Lithodomi*, and the stratification is as irregular as running water could make it. The occurrence of the oyster-shells made it doubtful if this deposit did not belong to an after-stage of the coral reef, distinct from these strata, because there is on the top of nearly every limestone cliff, which has not suffered much from denudation, a bed of oyster-shells united with a few other fossils, such as the asterite, and, what is not singular, considering the date of the beds, clearly established by other fauna, a *Pecten Jacobæus*. It is not with chalk but a ferruginous yellow clay. It is seen at Portland (under the trap), and therefore more ancient than the upper crag at the top of the Murray cliffs, and in many other places on the coast and inland. But the oyster-shell bed at Mount Gambier really belongs to a part of the upper deposit; and we must therefore conclude that that fossil extended to both beds. I should think, however, that the oyster-shell bed is more recent than the coralline rock, not only because it is always found above, but because the fauna is so distinct. Probably it was formed previous to the commencement of deposition from the ocean current, but when the coral was still subsiding, and a deep sea over it. It may safely be said that the deposit must have extended over a large area, for its remains are distributed at intervals about the South-eastern District.

We have now before us, therefore, a series of remains which point very decidedly to the existence, after the death of the coral, of an immense sea bottom, covered by deposition from an ocean stream. Before any conclusions are drawn from the facts stated, several peculiarities in the rock must be mentioned.

In the first place, the formation is one which, from the description I have already given, is perceived to vary much in its capability for resisting the action of sea-water and the atmosphere. The consequence is, that while some parts are easily washed away by coast action, others become compact and indurated. This, as the upheaval of the bed has proceeded, has given rise to reefs of rocks far out to sea, the more dangerous because they are rounded, and rarely visible above water. The coast from Rivoli Bay to Guichen Bay is very perilous to navigators in consequence, and a large reef of rocks to the north side of the latter (Cape Jaffa reef) stretches out to sea for more than twelve miles.

The appearance of these rocks is very peculiar. Of some only a small pinnacle is spared, which raises itself above the waves like a channel lighthouse; others, again, have been a mass of table rock, through which the sea has worn many passages, giving them the appearance of bridges; and lastly, they cluster together like a group of islands, with flat tops and precipitous sides. The flatness of the summits shows that they have been much denuded before arriving at their present state; but even amid their rounded form and worn outline the cross stratification is still traceable. The nearer they are to the coast the more rugged they become, until the rocks which fringe the shore are as studded with points and projections as a Gothic pinnacle or a melting glacier. In fact, their tops have just the appearance of a coral reef, quite as delicately branched, and as varied. A mere description can scarcely do justice to the strange appearance they present. It seems as if the rocks were covered with slender stone shrubs, tapering gradually to a point, amid numerous branchlets and ramifications, or as if the roof of a cave studded with stalactites were turned upside down, and placed on the seacoast. Anything but spray must long ago have broken them to pieces; and even then, how they have been spared, while the surrounding rock has been worn away, does not appear very plain.

Their origin I explain thus:—It would appear to me that they must be the result of concretions of lime and sand, caused by the percolating of water through the beds prior to their upheaval. This would harden some portions, and enable them better to resist the action of water. Indeed, the fact is evident that such a course has been in operation in other places.

It will be remembered that, in the second chapter of this work, reference is made to a kindred circumstance giving rise to concretions in the coral rock. At a cliff at Guichen Bay, out of reach of the sea, where portions of the rock have fallen away and caused cavities, the sides are seen to be covered with what appear to be roots of trees. Some are thick, and twisted in various forms, more divided at the top and thicker at the bottom, while others are slender. There is some difference between them and the ones now alluded to in this chapter. While the former are large and generally covered with a fragmentary mass of shells, the latter are small, and like chert inside, and covered with the white powdery chalk outside.

The same peculiarity may be seen at Cape Grant, already alluded to, where, as described, the summit overhangs the lower part of the face of the cliff, which is seen covered with these twisted concretions, of all sizes and sometimes continuous through many strata. Occasionally they jut out from the face of the rock, so as to be easily mistaken for roots. I attribute their origin simply to the filtration of water through the loose texture of the crag, which has dissolved the lime as it has passed through, and redeposited it round the channel it formed for itself.

There are, however, other theories extant as to the origin of similar beds, to which I must refer in vindication of my own opinion. The similar appearances have been noticed by two persons, both eminently qualified to give an opinion on such a matter,—by Mr. Charles Darwin, in his 'Voyage of the Beagle,' and by Mr. Gregory, in his account of the 'Exploration of Western Australia.' It is with the greatest diffidence I put forward my views (apparently in opposition to theirs, but not so in fact), that readers can see the statements and judge for themselves. I may, however, state, that my theory is the result of a longer consideration of the locality I refer to than either gentlemen were able to afford it, otherwise they would perhaps have seen the truth of what is here stated. It must be remarked, that I am not sure that Mr. Darwin refers to the same rocks as I do, and, from his statement and description, it is probable he does not; but as he himself alludes to other portions of the coast, very likely it may be imagined that his description was meant to account for all, especially as they are so similar. I do not doubt that the deposits of Western Australia were subaerial, because of the land-shells, though I should be more inclined to think that the casts were entirely due to the percolation of water. But I am sure that, had Mr. Darwin seen the deposits of Guichen Bay and Portland after those of Bald Head, he would very likely have set them down as identical. It is because the strata are so very similar that I am anxious to point out the difference.

In describing the rocks about Western Australia, he says:—'One day I accompanied Captain Fitzroy to Bald Head, the place mentioned by so many navigators, where some imagined they saw corals, and others that they saw petrified trees standing in the position in which they had grown. According to our view, the beds have been formed by the wind having heaped up fine sand, composed of minute particles of shells and corals, during which process branches and roots of trees, together with many land-shells, became enclosed. The whole then became consolidated by the percolation of calcareous matter; and the cylindrical cavities left by the decaying of the wood were thus also filled up with a hard pseudo-stalactrial stone. The weather is now wearing away the softer parts, and, in consequence, the hard casts of the roots and branches of the trees project above the surface, and, in a singularly deceptive manner, resemble the stumps of a dead thicket.'

Now, the only particular in which this description

differs from what I am describing, is the occurrence of land-shells; these I have never met except in the sand on the surface above the rock, which sand was evidently derived from drift, being composed of larger fragments of shells, in which the colouring matter was nearly always preserved. The land-shells were small species of *Succinea*. There were also in this drift many twisted roots and branches of mangroves and of other salsolaceous shrubs. These, if the deposit gets hardened, may possibly become like the roots at Bald Head; but then they will present a very different appearance from those described in the strata underneath. If the existence of land-shells in the Bald Head beds is as certain as their absence from those in South Australia, then we must clearly be treating of different formations. At any rate, lest anyone should imagine that our upper crag and its concretions owe their origin to a similar cause, it will be useful to state the reasons against such a theory. First, however, it must be mentioned that Mr. Gregory, the explorer, alludes to the same formation, and adds that it is derived from a drift of sand and shells from the coast which becomes hardened; and further, that it may be seen in all stages of formation round the coast. Probably he alludes also to our part of the coast; and this, again, is why I am anxious to state the difference.

It is true that sand is being drifted up in immense quantities round parts of the coast of Southern Australia, so as to bury trees and render considerable tracts unavailable. But this drift is only composed of the finest particles of shells and quartz,— such, in fact, as would only be carried along by winds. I have examined many of these 'sand dunes,' and a lengthened description of them will be found in the next chapter. It will only be necessary to mention that they contain no perfect shells, as far as I am aware, and never bear signs of stratification. Such a formation could scarcely be hardened except by the permanent action of water. Mere rain would not do it, and in fact does not; for after the winter season we find these banks as shifting and as loose as ever.

Then, with regard to the concretions, whatever may be the case with those in Western Australia, nothing but a very superficial observation would bear out the notion that they have ever been trees or roots; though they certainly have a strong resemblance in their roundness, and in the inequalities of the surface which give the appearance of bark: for, on being traced down, they generally continue for twenty feet without a change in their diameter, unless to become a little wider. Again, most of them at some part of their course get accessions from other percolations, and then go down in the form of fluted columns, which is hardly consistent with the notion of their being casts of trees. One would certainly expect to find also some trace of their vegetable origin, even though they be casts and not silicified trees, but nothing of the kind is seen. On breaking them, the interior is found to be a compact magnesian limestone, just what the filtration of water holding lime and magnesia in solution would occasion.

I have been often taken to see what have been termed fossil trees in the crag, but have always returned disappointed. Sometimes persons have shown me circular holes, about a foot in diameter, lined with concentric rings of limestone, and I have been asked, did I not consider them to be casts left by the trees which have rotted away? But, however delusive the appearances were, a reference to the sea-coast showed the holes to be analogous to the 'sand-pipes' spoken of in a former chapter. Near the sea they may be noticed of various depths, from one foot to five, and even more. They are always lined with concentric laminæ of stone. But the clearest proof that these strata have been formed under water is given by their present distribution. If the theory of Messrs. Darwin and Gregory were applicable to the upper crag of South Australia, then it must have been formed on dry land. But many portions of it are now under the sea, and consequently there must have been subsidence since its formation to place it there. Now, it may be safely affirmed that there is no evidence of such subsidence; on the contrary, a regular course of upheaval since the formation of the coralline beds is manifestly apparent. Therefore we may reasonably conclude that the theory will not meet the case of South Australia.

Some of the concretions may have resulted from

coral, transformed by rolling and coast action into rounded cylinders of calcareous fragments, and then buried in the fragmentary detritus. Mr. Darwin mentions such things as being common at Keeling Atoll, and I have frequently picked up upon the coast similar specimens.

It may seem a waste of time to dwell so long on a comparatively unimportant point; but I have been constrained to do so, not for the sake of contradicting abler men than myself, who were probably quite right in the strata they referred to, but in order to give an accurate account of the origin of all the beds met with in the district I have undertaken to describe.

After having described the concretions which make the coast more beautiful and less monotonous than the South Australian coast usually is, other peculiarities in the same strata soon claim our notice. The soft nature of this rock has already frequently been remarked. This has given rise to caves of various depths; and the rocks are so corroded, that one may wander long at Guichen Bay without exploring all the winding passages and crevices in the cliffs. Nearly all of them are more or less undermined, and scarcely a year passes in which huge masses of rock do not fall down. In some places, where the sea has been beating away at the end of a cave without having had much effect upon the sides, the water has bored to the surface by a kind of channel, through which every wave which falls upon the shore sends up a column of water into the air. These are the celebrated 'blow-holes,' of which there are many round the coast. Nothing can be more singular than their effect, even on the calmest day. You stand at the edge of a small round hole surrounded with stony shrubs, and every few minutes a roar is heard as the wave advances up the cave: it drives the air before it, and amid the noise a final explosion shakes the ground, dashing a cloud of silvery spray into the air, after which the water falls splashing around; there is a moaning recoil of the water, and stillness returns.

It is curious to observe the effect of two antagonistic forces which are here at work. The land is upheaving slowly, and the sea is rapidly eating away the coast line; there can be little doubt that the ocean would have the best of the struggle, and soon indemnify itself for all the losses made by the uplifting of her ancient bed, if the rock were all of the same soft texture. But this is not the case, as previously stated; and while the coast action has eaten deeply into the line of cliffs, causing either deep indentations or piles of ruins, some portions are able to hold out in the form of the reefs mentioned. The coast suffers much more heavily in winter than in summer; for it is quite unprotected from the whole southern ocean, and when the west and southerly gales prevail in winter, one day's wind is sufficient to send the sea upon the rocks with a fearful swell, which bears down everything before it.

We might further enquire, What has become of the detritus caused by these immense ravages? A great deal of it, certainly, is drifted up in the form of sand, which at Guichen Bay and Cape Bridgewater forms low hills, extending to the northward, and sometimes far inland. A portion of it, however, is redeposited on the tops of these rocks, which are under water, and forms a thin stratum of limestone, containing shells of existing species. The description of these beds belongs to the next chapter; but I may mention that they are found distributed, more or less, over the whole district. I have been the more anxious to dwell on the waste that the rocks are undergoing, because it will be presently seen that a great portion of them has been entirely removed; and unless we were previously aware of the way it is being worn down by the sea, we should have a difficulty, even taking into consideration its soft nature, in attributing it all to denudation.

Let us now go through the history of the deposit as the rocks present it to us. While the coral was yet building, the land continued to subside, and reefs which had been close to the shore became farther and farther removed from it. By and by coasts became islands, becoming smaller and smaller as they went down, and in time little more than a ring of coral was seen to mark the spot where they had been. Island after island disappeared, until at last the coral stood alone, a series of atolls and long reefs, in a deep and open sea. Changes in the relative position of land and water would give rise to changes of temperature, and this would in time cause new currents to flow in the ocean. Such currents bring down sediment, or fine deposit of broken shells, sand, and mud. Those corals which had not been killed by the alteration of temperature, or died by the subsidence or gradual diminution of the size of their reefs, would be destroyed by the sediment. Perhaps these are not the only changes to which their extinction is attributable. Changes in meteorological conditions would be so numerous after the subsidence of a large continent, that we could not say how many might cooperate to produce the same result. Thus, a change of wind for any length of time *causes the destruction* of many, by taking them out of the reach of the water. Besides, we do not as yet know all the conditions of animal life well enough to assert that races have not their duration of life as well as individuals. One thing is certain, that in going through the past history, as afforded by geology, we constantly meet with evidences of the destruction of a whole race by apparently natural causes. Bones are found in immense numbers round lakes; fossils which are plentiful in one stratum are quite absent from those immediately above; and many other instances might be given, all showing that a termination may come to animal existence without any apparently extraneous causes. Even something analogous may be seen in the human race, where new diseases are constantly appearing, sweeping away thousands of our fellow-creatures, and perhaps limiting the duration of human life.

At any rate, we have evidence of two things in the strata before us,— the death of the coral, and its burial under a large deposit of sediment. The subsidence seems to have continued long after the reef-building had stopped, and long enough, indeed, for the formation of the immense deposits of calcareous sandstone which now lie about the coast. Some parts of the coral reef are covered to a depth of sixty and seventy feet with the upper crag, all formed of the thin diversely stratified sandstone. Though the quantity of matter thus transported must have been enormous, it did not necessarily take such a long period to form as one would at first be inclined to think. Currents sometimes bear down an immense quantity of sediment in a very short time, and though this gets distributed over a large surface, yet it would not require very many years to cover a considerable area to some depth. Thus, then, it was that the upper crag was formed. The sea rolled over the reefs, carrying fragments of shells and sand from shallow places; the white mud was gradually darkened with the detritus it spread thicker and thicker, now in sandbanks and undulating hillocks, then a change would come; a stream from another direction, perhaps, which would sweep a level surface before depositing fresh material, leaving a deep line of division in the stone which was hardening underneath. Thus it went on forming either in huge mounds or level surfaces, small pieces of shell and fragments of sand adding and adding their tiny proportions to the mass, until the work was done, beautified by the percolation of water, and the stone was raised from the sea, as we see it now.

The above is the description of the way the stone was formed: we have yet to examine the evidence of its partial destruction. It may have been that the whole district, perhaps as far east as Port Fairy, to the Murray mouth west, has been covered with the same deposit, extending far inland, and that afterwards the greater portion has been removed, as the land was slowly raised from the sea. This opinion is founded on the occurrence at various parts of the country of small hillocks of rock, perhaps an acre or so in extent, and some few feet thick, identical in composition with the upper crag, even to the concretions. There are such spots at Mount Gambier, and again at many parts of the Mosquito Plains. These are generally very hard indeed, like granite, are much rounded by the sea, and are generally perforated all over by *Lithodomi* or molluscs, which bore into stone lying under the sea. Such borings show that the stone itself has been exposed for a long time to coast action, and its hardness explains how it was able to resist decay and withstand wear, which swept away the rest of the deposit.

Of course it cannot be said that the upper crag was at one time spread over the coral rock in the same thickness as the strata seen at Guichen Bay and elsewhere; the continued appearance of such patches as those just described show clearly that it was pretty generally distributed; but it is of the nature of such strata to be partial, and to have heavy banks in some places, while in others it is nearly entirely wanting: neither at Cape Grant nor at Guichen Bay are the strata of the same thickness throughout. Sometimes they are piled up like sand-banks, and at other places they are low. At Cape Lannes, before alluded to, the rock is very hard, while a narrow neck of the same strata which joins it to the shore is comparatively softer, and is rapidly being worn away; so that the cape may yet stand out as an isolated rock. From this point is also observed the varied hardness of the stone; the coast behind is honeycombed into the most fantastic forms, bearing no small resemblance to ruined Druidical monuments amid ornamental garden pottery.

In a rock where hardness is the exception rather than the rule, it is not difficult to imagine how such immense portions became removed. As the upheaval went on, each separate portion was exposed to the action of the coast, which not only appears to have perfectly removed it, but to have eaten for some distance into the coralline rock underneath. The denudation thus effected must have been enormous, even admitting that the deposit was not general nor of great thickness throughout; and much as it excites our admiration to see millions of tons of rock brought down by an ocean stream, still more are we astounded to see the same swept away again so completely, that, had not a few traces remained here and there, we could not have known even of the existence of what took ages to form.

When such operations as these are brought to light, we perceive the utter impossibility of arriving at any correct data for a geological chronology. Whole strata may have been removed, and the detritus stratified elsewhere, and this again denuded; while not a vestige of these operations remains, and the immense time occupied in their accomplishment remains wholly unknown to us. Thus it is again that such singular breaks occur in the geological history of the earth in going from one period to another. A totally different fauna will succeed within a few feet. The record of the changes gone through has probably been denuded, and left us with only the hiatus evident. *Causa latet, vis est notissima.*

Before quitting the subject of their formation, there are two or three things yet to be noticed. The age of these rocks — that is, their position in the geological series — is easily determined; for though there are no fossils, there is the record of these strata immediately following the coral beds in their subsidence: above, again, are more recent strata, to be made the subject of the next chapter. These may here indicate a formation which has a strong resemblance to our crag, namely, the Suffolk crag, described by Charlesworth, Phillips, Lyell, &c. I will here repeat what has been said in a former chapter:—' The Suffolk crag is divisible into two masses, the upper of which has been termed the Red, and the lower the Coralline crag; the upper deposit consists chiefly of quartzose sand, with an occasional intermixture of shells for the most part rotted and sometimes comminuted. . . . The lower or Coralline crag is of very limited extent, ranging over an area about twenty miles in length and three or four in breadth. It is generally calcareous and marly; a mass of shells, bryozoa and small corals, passing occasionally into a soft building stone . . . . At some places in the neighbourhood the softer mass is divided by thin flags of hard limestone, and corals placed in the *upright position* in which they grew. The Red crag is distinguished by the deep ferruginous or *ochreous colour of its sands* and fossils, the Coralline by its *white colours.*' A little farther on, the same author says:—' The Red crag, being formed in a shallower sea, often resembles in structure a shifting sandbank, its layers being inclined diagonally, and the places of stratification being sometimes directed, in the same quarry, to the four cardinal points of the compass, as at Birtley.'

Now this description would apply very well to our crag. The colour, the stratification, the irregularly deposited comminuted particles resting on a white coralline rock, would seem to be identical with the deposits treated of in this and the former chapter. But mineral composition alone is a very weak guide in these matters. It is upon the fauna we must depend; and in this, as far as I can judge, they are not dissimilar. But still it is interesting to observe that a coralline rock, like that of Australia, has been followed by a sandy deposit like our crag.

That similar fossils should be deposited under different local circumstances, so as to have an almost entirely different mineral character, is not at all uncommon; but that dissimilar strata, containing fauna, related to each other by a similar geological epoch, should be deposited under precisely similar circumstances, is a remarkable instance, as stated in last chapter. Professor Forbes concluded that the Suffolk crag was not found at any great depth of sea, probably at not more than from twenty-five to thirty fathoms; but yet he would not call the deposit a lateral one, because it might have been fifty miles out to sea.

The same might, perhaps, have been said of our crag, did it contain any fossils. But as there are none, and as the portions of shells are all very finely broken, perhaps the depth of the sea and the distance from land were much greater than that of the Suffolk crag. The sea might have been in course of upheaval during the formation of some parts of it. I should imagine this from the layer of trap upon which a portion is stratified. Volcanic emanations are only usually met with when the land is uprising. But this subject more properly belongs to another part of this work which treats of the volcanoes.

In conclusion, I must mention a few facts with

reference to the water-level here. Generally fresh water is found on the top of the crag when covered by more modern strata. If it is not found there it must be sunk for until the compact portion of the coralline rock is reached. At Mount Gambier the water seems to preserve a uniform height above the sea, so that the depth of a well will depend upon whether it is sunk on rising or low ground. The water is very hard, and contains large quantities of magnesia. It is singularly clear and pellucid, but, when more than forty feet in depth, it exhibits as rich a blue as the deepest parts of the ocean.

The colour of water will very often depend on the bottom upon which it rests. Thus I have seen the sea a light green when out of soundings, and many hundred miles from the African coast; and I have seen the sea a deep blue at thirty fathoms, close to a basaltic coast: but at Mount Gambier it appears to be the nature of the water; for, no matter how white the limestone beneath may be, the water if of any depth is deep blue. I believe Bunsen has published reasons why blue is the natural colour of water; but, I think, if a careful examination were made, its colour would be found to depend upon the salts it holds in solution.

I may add here, that when approaching the Australian coast, we took numerous soundings, and found that at about ninety fathoms (540 feet) the lead came up covered with a loose fine deposit of broken shells and sand exactly like the crag; this might give some idea of the depth at which such strata might be found. I have found no *Foraminifera*, though doubtless they exist in the deposit; under the microscope, it appeared as if composed entirely of small fragments of shells.

CHAPTER VII.

THE REEF'S SUBSEQUENT HISTORY.

PRELIMINARY OBSERVATIONS.—ASPECT OF THE AUSTRALIAN COAST.—SAND.—SAND FORMATION OF CORNWALL.—ORIGIN OF AUSTRALIAN SAND.—ITS COMPOSITION.—UPPER LIMESTONE AND SHELL DEPOSIT.—LOCALITIES IN WHICH THE LATTER OCCUR.—STONE HUT RANGE.—OBSERVATIONS ON THE FAUNA OF THE DEPOSIT.—LAKES ON THE COAST.—THE COORONG.—LAKE HAWDON.—LAKE ELIZA.—LAKE ST. CLAIR.—LAKE GEORGE.—LAKE BONNEY.—GERMAN FLAT.—MOUTH OF THE MURRAY.—UPHEAVAL OF THE AUSTRALIAN COAST.—THIS PROVED FROM THE COAST LINE—FROM SOUTH AUSTRALIAN RIVERS, AND ESPECIALLY THE REEDY CREEK.—UPHEAVAL STILL GOING ON.—PERIODS OF REST.—SIX CHAINS OF HILLS.—TERRACES FORMED FROM OLD SEA BEACHES.—SAND DUNES NOT HARDENING INTO STONE.—SIMILAR FORMATIONS IN SUFFOLK.—LAKE SUPERIOR AND BAHIA BLANCA.—WHY GENERALLY ASSOCIATED WITH SANDSTONE.

WE have hitherto been considering the underlying rocks of the districts. Though they often crop out, and are always met with only at a very few feet from the surface, yet they may properly be termed underlying, because they are geologically more ancient than the deposits to be considered in this chapter. We have seen just now how the whole series of strata, from the coralline to the crag, resembles the series that is found in Suffolk, as far as mineral composition and general aspect are concerned. This correspondence is the less singular, once that the coral reef theory is admitted, for we find that round most atolls, barren reefs, &c., the broken coral generally becomes formed into a hard rock, like ferruginous sandstone, which is very compact, though composed of large grains. Perhaps some patches of crag found in the interior were formed thus, contemporaneously with the coral, and are therefore distinct from the crag of the coast: but it must not be thought that the whole is of the same origin; its thickness, its stratification, and its general texture preclude such a supposition.

We are now about to consider a deposit with which coral has had nothing to do; it is neither so extensive nor so thick as the others we have been considering, and it is the last in the geological chronology of this part of Australia.

Its description must necessarily be rather lengthy. There are so many features, so many details to be considered with it, that even a mere enumeration would be long. The details, however, are mostly of an interesting character, and their consideration will repay the importance allotted to them.

As we have now to deal with the surface, let us begin with the coast line. The aspect of the coast of Australia, like that of Egypt, Arabia, and many other countries, is low and sandy. Places here and there, like Cape Otway, Cape Jervis, and Port Jackson, besides other small spots, expose a bold and rocky front to the sea, but generally only sand-hills are seen, dotted here and there with dark-green patches, but commonly forlorn and uninviting.

It appears strange how early discoverers could entertain a good opinion of the country, when all that met their view was an interminable line of sand and scrub, rendered unapproachable not only by its cheerless loneliness, but also by the large white surfs which boomed eternally along it with a gloomy roar. Hills and trees might appear in the distance, but really they are the exception on the east coast, and even these are thickly wooded with vegetation which is anything but verdant or promising.

Let us commence, then, with the sand. This covers the shores, not for a few hundred yards, as on most beaches, but sometimes a mile or more inland. Seldom are rocks seen amongst these hills; if there are any, the sand covers them,—not a fine white silicious sand, but coarse grained, containing many small fragments of shells, of a light-yellow colour, and composed nearly entirely of carbonate of lime. It has been mentioned in the second chapter of this work, that a great part of the district now described is covered with a calcareous sand. As we know that the whole coast has been upheaved, perhaps some of it has been derived from coast action. At all events, for twelve miles inland sand is common, even though covered with grass; and that this has been sea-sand there can be little doubt, because it is interspersed with sea-shells such as are now found upon the coast. Considering its calcareous nature, it is a matter of astonishment that it has not been consolidated into a compact rock. But it has not been. On the contrary, it is so loose as to be blown about by every breath of wind, giving rise to the phenomena of sand dunes (like those on the Suffolk coast), which will be more fully noticed when I come to speak of upheaval.

Such sand might, however, be hardened into a stone. Instances of this are met with at Guadaloupe and on the coast of Cornwall. The latter is worth citing here, more especially as it shows how a subaerial deposit might resemble the crag mentioned in the preceding chapter. It is thus described in the appendix to Mantell's 'Wonders of Geology:'—

'A sandstone occurs in various parts of the northern coast of Cornwall which affords a most instructive example of a recent formation, since we here actually detect nature at work, in converting loose sand into solid rock. A very considerable portion of the northern coast of Cornwall is covered with calcareous sand, consisting of minute particles of comminuted shells, and in some places has accumulated in quantities so great as to have formed hills of from forty to sixty feet in elevation.

'In digging into these sand-hills, or upon the occasional removal of some part of them by the winds, the remains of houses may be seen; and in places where churchyards have been overwhelmed, a great number of human bones may be found. The sand is supposed to have been originally brought from the sea by hurricanes, probably at a remote period.

'At the present moment, the progress of its incursion is arrested by the growth of *Arundo arenacea*. The sand first appears in a slight but increasing state of aggregation on several parts of the shore in the Bay of St. Ives; but on approaching the Groythian river it becomes more extensive and indurated. On the shore, opposite Godrevy Island, an immense mass of it occurs, of more than a hundred feet in length, and from ten to twenty feet in depth, containing entire shells and fragments of clay-slate; it is singular that the whole mass assumes a striking appearance of stratification. In some places it appears that attempts have been made to separate it, probably for the purpose of building, for several old houses in Groythian are built of it.

'The rocks in the vicinity of this recent formation in the Bay of St. Ives are greenstone and clay-slate, alternating with each other. The clay-slate is in a state of rapid decomposition, in consequence of which large masses of the hornblende rock have fallen in various directions and given a singular character of picturesque rudeness to the scene. This is remarkable in the rocks which constitute Godrevy Island. It is around the promontory of New Kaye that the most extensive formation of sandstone takes place.

'Here it may be seen in different stages of induration, from a state in which it is too feeble to be detected from the rock upon which it reposes, to hardness so considerable that it requires a very violent blow from a sledge to break it. Buildings are here constructed of it; the church of Cranstock is entirely built with it; and it is also employed for various articles of domestic and agricultural uses.

'The geologist, who has previously examined the celebrated specimen from Guadaloupe, will be struck with the great analogy which it bears to this formation. Suspecting that masses might be found containing human bones, if a diligent search were made in the vicinity of those cemeteries which have been overwhelmed, I made some investigations in those spots, but, I regret to add, without success.

'The rocks upon which the sandstone reposes are alternations of clay, slate, and slatey limestone. The inclination of the beds is SSW., and at an angle of 40°. Upon a plain formed by the edges of these strata lies a horizontal bed of rounded pebbles, cemented together by the sandstone which is deposited immediately above them, forming a bed of from ten to twelve feet in thickness, containing fragments of slate and entire shells, and exhibiting the same appearance of stratification as that noticed in St. Ives' Bay.

'Above this sandstone lie immense heaps of drifted sand. But it is on the western side of the promontory of New Kaye, in Fishel Bay, that the geologist will be most struck with this formation, for here no other rock is in sight. The cliffs, which are high, and extend for several miles, are entirely composed of it; they are occasionally intersected by veins and dykes of breccia. In the cavities, calcareous stalactites of rude appearance, opaque, and of a grey colour, hang suspended. The beach is covered with disjointed fragments, which have been detached from the cliffs above, and many of which weigh two or three tons.'*

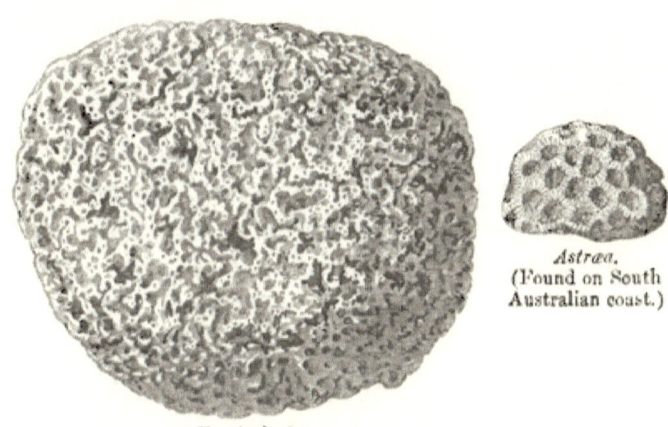

Fascicularia.
(Now found on South Australian coast.)

Astræa.
(Found on South Australian coast.)

The sand of our Australian coast appears to have been washed up from the sea, and not derived from a hurricane. It is important to enquire how; for though sand on coasts is a very ordinary thing, yet the large quantities of it here are worth some attention. Some must be derived from the rocks that are being washed away, and the rest from the shells and corals which frequent the shore. There is an *Astræa* rather common on the coast, and some Nul-

* From a Memoir by Dr. Paris, in the *Transactions of the Royal Geological Society of Cornwall.*

lipores exceedingly so. Corallines are also very often met with, including probably the rare *Fascicularia*.* The fragments of these, broken small by the beach surf, raise the mounds of sand in such small quantities that it soon dries and remains light without consolidating: the wind carries it farther inland.

It is to be remarked, that a coral shore always gives rise to a calcareous sand. Thus, in the atolls, the sea breaks upon the living coral, and then spreads over a sort of terrace composed of the hardened calcareous sandstone before alluded to. Finally, it washes up a belt of sand upon which the cocoa-nut palm and shrubs grow. The sand-hills thus raised are described as being loose, white, and calcareous, seldom rising to more than twelve feet above the level of the sea. The banks on the Australian coast do not rise very high. Occasionally, indeed, a mound will attain the height of 150 feet, or sand may drift until it forms a slope of even much greater altitude, but the general nature of such formations is low and even. .

This sand, where it accumulates on large sand drives, is very yellow in colour, and seems to consist almost entirely of the broken fragments of shells. On most parts of the coast it is white and rather fine. Under the microscope this latter is very in-

---

* The *Fascicularia* was thought to be peculiar to the Crag, and a characteristic fossil. The one engraved is common enough on the coast, though, curiously enough, it is not found as a fossil in the Mount Gambier limestone. It must be said, however, that it does not correspond in every particular with the *Fascicularia* of the Crag.

teresting. It teems with the remains of animal life. First, the spines of *Asteridæ* and *Echini* are easily separated; then come fragments of *Bryozoa* and shells. After this, *Foraminifera* are traced, sometimes of a size sufficiently large to be picked out with the help of a small pocket-lens, sometimes so minute as to be with the utmost difficulty transferred to the microscope; but the most common (so much so, that the sand may be said to consist principally of it) are sponge spicula, as clear as glass and of every shape, but mostly circular and triradiate. The proportion of mere inorganic silex is very small, so that these immense masses of sand which belt the coast for thousands of miles may be regarded as a mass of microscopic organisation. How many millions of animals must they not contain!

Thus far one feature of the coast. It has been said just now that limestone is uncommon on these hills; and indeed, wherever it is seen, it is only in breaks in the banks which cover it: but at a moderate depth, and even on the tops of the hills, for twelve miles inland, on the crag sometimes, and composing hill-sides itself in some places, a smooth white limestone is found, very compact and pure. This is another formation which overlays the upper crag. From the crag it may be distinguished by its never being granular, and from the coralline by being more compact, and of a dull yellow colour approaching to brown. It is never more than about three feet thick, sometimes not even so much. This makes it directly the opposite to the coralline rock, which is always (except where the strata thin out) of very considerable thickness; the difference is owing to the different circumstances under which they were formed. The coralline was deposited during a period of subsidence, when the strata had ample time and fauna to thicken it, and no disturbing cause. The Post-Pleiocene (as we shall henceforth term the deposit) was spread out over rocks which were continually being taken out of the reach of the waves by upheaval, and therefore could not attain any considerable thickness.

As it is, however, it is full of shells, but little altered in the composition from what they were when in the sea, sometimes even preserving their colours, and all of them of species that are now found on the coast. It is as well to observe here that this shelly limestone often overlays the upper crag, and therefore adds another argument against the subaerial nature of that deposit. The shells, though strictly identical with those of the coast, are not in the same proportionate numbers. Thus a person might dig for hours around the rock, and yet find nothing but a *Venus exalbata* and a *Cerithium*.* Now these shells, though common on the coast, are far from being the most common. What most frequently is found upon this part of the South Australian shore is a large

---

* Sometimes immense masses of rock are formed entirely by the shell represented in the engraving.

Trochus, also a *Conus* and a *Littorina*, and these are rarely met with in the strata. These facts show that some alteration has taken place in the coast fauna since the rocks were deposited.

The place where the deposit is seen to best advantage is on a range which, without exception, runs from two to twelve miles all round the coast, from the Coorong to within a few miles of Portland, in Victoria. But it is not confined to this range; sometimes between it and the sea there is another and less extensive line of hills, and then the space between them is a flat filled with salt and freshwater lakes. These flats and the hills between them and the sea are covered with the same recent shells, but there is no limestone except in patches here and there on the edges of the lakes or on some low hills, and the shells (for they can scarcely be called fossils) lie in loose sand or in thick beds with little or no sand. It is impossible to exaggerate the enormous quantities of shells which are sometimes seen on the flats. The roads are in consequence firm and dry, just like the shelly walks of a park, and when occasionally a strong wind tears up a tree by the roots, the fibres have become so interlaced with these spoils of the ocean as to look like some large article of fancy-work. On the coast, too, amid the sand-hills, shells are again seen, not lying here and there, as though brought by human agency, but regularly stratified into the mass, so as to show, when exposed, regular layers like strata.

It may be remarked, in passing, that this fact affords another proof that the crag cannot owe its origin to the hardening of drift sand, as sea-shells are never found whole in it, and even the fragments of such are not regularly stratified; but here we have sand clearly owing its origin to drift, with regular layers of sea-shells, thinly scattered, it is true, but showing that each portion which was exposed was liable to have sea-shells deposited upon it.

Returning now to the range parallel with the coast, sea-shells are found upon the flat at its back. How much farther inland they are seen I cannot precisely say, but they have been found by me embedded in limestone at least seventeen miles away from the sea. It will be remembered, that in describing the features of the south-eastern district, a series of ranges, running north and south, about twelve miles apart, with flats between, were spoken of. These will have to be referred to again when speaking of upheaval, as probably they have all formerly been coast lines. Now the flat at the back of the range nearest the coast is more or less covered with shells, as just stated, but they are found in a peculiar manner. The soil is soft black or whitish clay, and the limestone does not lie continuously underneath, but here and there are patches 'caked' into the surface, just as if it had dried on the margin of a lake. The fossils, too, are of a mingled description, being partly fresh water and partly marine.

Now this flat was in all probability an estuary,

because at one place, where the fresh-water shells are commonest, there is a gap in the range about six miles in width. This gap is at present occupied by an immense morass called Lake Hawdon, which is much longer than its width, and which is narrowest at the above-mentioned gap. There can be no doubt that the range (let us, for convenience, call it the Stone Hut Range, by which name it is known by residents) has, until very recently, been a coast range. The shells on its summit, the shell banks to the west of it, and the smooth even way it is washed on the coast side, besides having the stone perforated by the *Lithodomi* on the same side, all declare this. The gap then would allow an opening to the flat behind, which would be alternately filled with salt and brackish water, for the quantity of fresh water which settles on these flats in the rainy season would materially affect an inlet which only received occasional accessions from the sea, and this we may, from the shallowness of the gap, suppose to have been the case.

Traces of recent shells have been found, more or less, on all the ranges to a distance of fifty miles from the coast; farther than this they do not, however, extend, and the country becomes volcanic. We will now confine ourselves to those localities nearer the coast, where the shells are loose and numerous, and attended with peculiarities which will demand some lengthened consideration. Before doing so, there are a few remarks to be made with reference to the deposits where the shells are embedded in the limestone. It can hardly be called a formation, because as yet it is not completed, and the sea is even now stratifying the same fauna into limestone on the coast. It is a thing of the present rather than the past. Although I have examined several hundred specimens, I have not found one that does not exist at present on the coast. It is true, as before stated, that some species are not so common now in the same localities as they were when parts of the limestone were forming, but there is no difference between them.

The following is a list of the most common genera now found at Guichen Bay. Those marked with an asterisk are very common in the limestone, and those marked with a cross are most common on the coast :—+ *Purpura*, *Fasciolaria*, *Haliotis*, + *Turbo*, + *Conus*, *Bulla*, *Ampulla*, *Natica*, *Pectunculus*, *Hyponix*, * *Venus*, *Iridina*, *Nerita*, *Panopœa*, *Pleurotoma*, *Fissurella*, *Cerithium*, *Turritella*, *Cypræa*, *Nassa*, *Trochus*, * *Phasianella*, * *Voluta*, *Mactra*, *Donax*, *Ostrœa*. Very many of these are common to both, and, though none are present in the limestone and absent on the coast, some are absent in the limestone which are very common on the coast. The limestone in which they are found makes excellent lime, and is a good and durable building stone, easily dressed.

The coast is the locality where the circumstances under which the deposit has been formed are best understood. A reference to the accompanying map will show better than a description how the whole of this part of the South Australian coast is covered with lakes and inlets, not running up into the interior, as in a mountainous country, but having their greatest lengths parallel with the coast. There is the Coorong, which is an arm of the sea, having its opening not very far from the Murray Mouth, running parallel with the sea in a narrow steep for miles, and terminating in a little creek which runs some distance inland. There is only a narrow strip of sand-hills between this singular piece of water and the sea for the whole distance; in fact, it looks more like a fringing reef to the coast than anything else. Next comes a little series of lakes, which are hardly worth mentioning, any more than that they bear out on a small scale the prevailing character of this coast. Next comes Lake Hawdon, which, as I have said, is more a morass than a lake, and which is an exception to the general rule, of the greatest length being parallel with the coast line; but, as it once evidently occupied a part of the flat behind the Stone Hut Range, and probably extended a long way behind, this exception is more apparent than real.

Next in succession comes Lake Eliza, a fine sheet of salt water, very shallow, and rapidly drying up. This latter lake is separated by a very small strip of land from Lake St. Clair, which is smaller than the former, but possesses much the same features. Then there is Lake George, an irregularly-formed sheet of fresh water, whose banks are reedy and muddy. It has two or three fresh-water creeks leading into it, but there is no apparent outlet between it and the sea. Finally, there is Lake Bonney, a long narrow sheet of fresh water, twenty-five miles long, but in few places more than about two miles broad; it is shallow, and surrounded on all sides by moderately high banks.

As all these lakes have distinct peculiarities, a separate description for each will be necessary, more especially the last one mentioned, which has so many various features as to be well worth attentive consideration. Let it be remarked, however, that though some of these lakes are marshy while others are clear and open, some fresh and some salt, they mostly lie in the flat which runs between the coast range and the sea. It may also be stated, that at Guichen Bay, there is a small range very close to the coast, and between this and the sea there exists a succession of small lagoons, some deep, with steep banks, while others are mere shallow pools. The water in them is fresh, or nearly so. Round their edges there is a stone in course of formation, which will account for the patches of limestone that are occasionally met with in the flats farther inland. From the appearance of the edges of these lagoons one would naturally conclude that they were salt, for they have all round a white crust just like salt. This crust is a sort of white lime and clay, quite hard and rough (enamellated like mollipora coral) for about an inch, but underneath soft and boggy to a considerable depth. It is full of shells, mostly marine, but some fresh-water, and contains abundance of a white chara or conferva, which grows plentifully on the lagoons. A person may in some places walk to the water's edge on the outer crust, but in others it is treacherous, and underneath the white sand is so deep as to render any submersion rather dangerous. Melaleuca and mangrove shrubs frequently grow on the edges, and their roots, &c., sometimes remain in the stone and mud beneath.

It can easily be seen how the study of localities such as these may throw a light upon peculiarities in the flats far removed from the sea. If, hereafter, upheaval should move these lagoons farther inland, they will dry up, and these large sheets of thin laminated limestone, enamellated or botryoidal on the surface, and containing shells — such limestone as now, in fact, is found farther from the sea — will be the result. The banks round these lagoons are not high, and in very rough weather the sea has been known to rise above the sand-hills which fringe the coast, and rush into them, thus making their character alternate between salt and fresh water.

To commence with the Coorong, a glance at its appearance as it is shown upon the map would lead one to believe that it has formerly been a fringing reef, whose corals have been destroyed by upheaval. Such is the aspect of fringing reefs in the Mauritius which have undergone that change, and they are described as mere sandbanks, which lie far out to sea, more like the earthwork of a fortification than anything else. I have never been able to give a sufficiently minute examination to affirm this positively, but I can see one or two objections to the hypothesis. Coral such as would form fringing reefs does not occur in any part of the south coast at present, neither has coral ever been found *in situ* on the Coorong itself. Again, a reef of this description must necessarily have given rise to a coral débris farther inland, and this also is wanting. Another more probable theory may, perhaps, explain this interesting geographical feature. It may have been a long sandbank under the sea, which has been raised by upheaval, while the intervening low land between it and the coast is as yet covered by water.

The basis of the bank may be a barrier reef of the crag period, which would serve as an obstruction upon which sand would gather, or it may have been an outcropping ridge of rock, but, at any rate, its texture and general features render the bank theory far more probable and consistent with its appearance. Some persons have imagined that it was the former bed of the Murray, but an inspection of its bank, as well as its south termination, will at once show that this view is quite untenable.

From the Coorong, the next lake of any importance to the south is Lake Hawdon. This has been already described. It may be added, that it is, with small exceptions, covered with long dense reeds, and that while at its western end sea-shells abound, at its eastern there are some fresh-water and sea-shells intermixed.

From Lake Hawdon to Lake Eliza is about six

miles. The latter is a good broad sheet of sea-green water, with a fine sandy bottom some miles long, and nowhere deeper than about eight feet. It is rapidly drying up. This is presumed from the banks, which are very flat for a long way round the edges, sometimes a mile, covered with black mud and the caked limestone already described when the lagoons were spoken of. Long before the water is reached, the banks become marshy, and are covered with a very dense thicket of melaleuca, callistemon, mangrove, &c., and the whole of the black mud, as well as the dry banks, are covered with sea-shells as thickly as the coast, and even more so.

Lake St. Clair is only divided from Lake Eliza by a narrow strip of land, which is sandy, and as thickly covered with shells as any other part. There can be no doubt that they formed one lake within a very recent period, because, supposing the water only to have extended as far as the black mud and shells round Lake Eliza showed that it did extend, there would have been quite sufficient to have covered the line of division between them. In that case, they would have formed one long strip of water smaller than the Coorong, but somewhat similar to Lake Bonney.

It may have been connected with the sea, supposing the water to have been much higher (for which supposition there is good reason), for though between one part of the lakes and the sea there is a high range of sand-hills and limestone ridges, at the northern end there is only a low marshy flat fringed with sand-hills, and running to one or two lakes which are more or less connected, until the sand-hills of the coast are reached. The soil on this flat is black, and supports a rank grass, but sea-shells are seen on the surface, and at a very small depth below they abound. If this was an arm of the sea, and Lakes Eliza and St. Clair were a deep bay, the Stone Hut Range, which rises to several hundred feet above the sea-level, must have been a beach, and never were signs more strikingly visible than those which still remain to give proof of the fact.

The hills rise abruptly from the beautiful level flat (except a gentle slope up to them, as in all beaches), and the limestone of which they are composed is washed smooth and clean, besides being perforated at the base by the borings of *Lithodomi*. There are ample marks of coast action on the stone, such as worn and weathered surface, deep circular hollows or wells, such as are found now on the coast, and lines with the same laminated limestone. There is also, at the foot, a deep deposit of sand and shells, broken and comminuted by beach action, and finally, what is most convincing, parts of the salt water of Lake Eliza still wash the foot of more southern portions. So evenly has the water cut off all projections, and so steep has it washed the approach to the flats, that there are only one or two places where a passable bush road can be formed through them.

Before leaving the subject of these lakes, it would be well to mention that they would have been dried up long ago, were it not for the existence of some fresh water, more particularly described by and by. At present, they only run in particular seasons of the year, and therefore are not able to counteract the immense evaporation which goes on, so that, eventually, the lakes may become perfectly dry, and give rise to black loamy flats with shells interspersed, such as are now met with farther inland.

Next in succession, proceeding southwards along the coast, we meet with Lake George. This is an irregularly-formed fresh-water lake, farther inland than these lakes usually are, and bounded, as usual, on its eastern side, by the continuation of the Stone Hut Range.

It has two or three fresh-water creeks leading into it, but none leading out. This is a peculiarity with more of these pieces of water; they all receive tributaries of some kind, but have no outlet between them and the sea, and this may be accounted for by the fact, that nearly if not all of the tributaries are dry during the summer and perhaps the greater portion of the year, and the amount of water gained by them is more than lost by evaporation. In fact, all the lakes bear some remote analogy to the Dead Sea in Palestine, which appears to have been an arm of the sea formerly, and though there is never any overflow into the Red Sea, it receives the whole waters of the Jordan. If evaporation, in this case, can consume the enormous supply of the Jordan, which is always flowing, the climate of this part of Australia, which is quite as dry and rather hotter than that of Palestine, can easily dispose of the result of a few weeks' rain. I have not been able to afford time for a personal examination of Lake George, but I am credibly informed that it is like most of the others, and has a deep flat of black mud thickly embedded with shells all round its edge.

Last of all in the series comes Lake Bonney, which, with the exception of the Coorong, is the most important of all. This, as before stated, is about twenty-five miles long and only about two broad, covering an area of rather more than fifty square miles. Like the others already described, it lies between coast hills which lie at the edge of the sea and a continuation of the Stone Hut Range. But these two almost seem to join at the north and south ends, or, at least, are nearly continuous, by a low line of sand-hills which lies between. The hills, however, on each side are almost exclusively sandy, and seem to rise to their greatest height opposite the middle of the lake. Round the water the appearances are very similar to what is observed elsewhere; that is, level flats covered with black mud, limestone, and salt-water shells.

The water is fresh or brackish, and very shallow. One or two creeks, as usual, flow into it, but there are few outlets. Evaporation goes on very rapidly, but it may be long before it dries up, for I have been assured that about ten years ago, after an unusually dry season, the greater part of the bed of the lakes was quite dry, so that persons have ridden across parts that are now completely under water. To such an extent as twenty-five miles of length, it might be expected that there would be great variations in the nature of the bank, and, accordingly, while some are flat and grassy, others are mere barren sand-hills, which rise rather abruptly from the bank, but there are no precipitous sides, nor, as far as I could learn, any rocks cropping out around its edge.

There is one peculiarity worthy of attention at the back of the hills which form the eastern or inland boundary of the lake; that is, a long swamp or marsh, which runs parallel with the hills for the whole length of the lake. This is called the German Flat, and is about twenty-five miles long and three broad. Here and there, places may be found where it is passable, but, in general, it is an immense quagmire, thickly covered over with dense reeds. The most superficial observation will convince any one that this has been a reservoir of water at a time when what is now the inland boundary of Lake Bonney was the coast line. Very likely, it was the drainage of the flat which lies on the eastern side of it (for there is no elevation of any note for at least ten miles inland), which congregated there when the sea had thrown up the hills, and probably this water occasionally received accessions from the sea in stormy weather, as in the case of the lagoon at Guichen Bay. There are, however, at present, no marine shells discoverable in it, and this might be expected, because the complete wall which now exists between it and Lake Bonney does not give room for the belief that there was any bay or inlet by which, as in the case of Lakes Eliza and Hawdon, the sea was admitted, until very recently, so that, had there been any shells, they would have had ample time for decomposition long before this, in the peculiar mud of the German Flat.

It is remarked that every year this flat gets drier, and the land, consequently, more available. This may be due to a greater dryness in the seasons than those which formerly prevailed — a fact to which all the older settlers bear testimony — or it may be due to an upheaval of the land. One thing, however, is certain, and that is, that persons who have been witness to the great changes which have taken place in the flat since their first location near it, confidently look forward to a time when the whole will be available for cultivation. The appearance of the flat when seen from Mount Muirhead, a hill about twelve miles distant, is that of a red-brown strip of land which lies like a desert round the hills of Lake Bonney.

Before leaving the subject of the lakes, two must be mentioned, which lie at the mouth of the river Murray: these are, Lake Alexandrina and Lake Albert. The former has evidently been a deep bay at the remote time when the Murray

Mouth was at its northern end. It is a shallow lake, like all of these, and, owing, perhaps, to the immense quantities of sediment which are brought down by the river, is becoming gradually more shallow. At the southern end, the Murray Mouth runs through it in a very narrow tortuous channel, which is constantly altering in depth, owing to the sand thrown up by the sea, which beats outside. Lake Albert is a piece of water adjoining Lake Alexandrina, and, like it, appears to have been a bay of the sea. It would appear as if both these lakes owed their origin to a cause like that which formed the Coorong. The upheaval has raised from the sea certain eminences which existed underneath the water as banks or shoal, and these being higher than the bottom between them and the shore, locked in the water as soon as they were above its level. Doubtless the hollow of the lake was caused by the river, and the sediment brought down by it may have caused the bank which, now being upheaved, forms its southern boundary. As usual in these cases, the banks of both lakes abound with existing species of marine shells, showing that all the operations which have taken place have done so within a recent geological period.

Upheaval of the Australian Coast.—It now remains to speak of that which has been so often alluded to in the foregoing chapter as a certain fact, namely, that of upheaval. After having shown that the whole coast round, to a distance of several miles inland, is covered with recent shells, and further, having shown that the drainage of the country is apparently altering, that the lakes known to have been formerly filled with salt water are now filling up with fresh, or becoming dry, it does not require any very great extent of argument to prove the upheaval of the land. But there are other facts. Let us pay attention to the coast-line first.

The mere outline of the coast seems to show what has taken place. The very fact of so many salt-water lakes near the shore which are not found inland, the majority of them being filled with salt or brackish water, and having their greatest lengths parallel with the coast, is just the state of things we can suppose as having arisen from a coast which the sea has left; and when we take into consideration that all the banks of these lakes are covered with marine shells, so recently derived from the sea as to preserve their colours in many cases, any doubt as to their recent recovery from the sea must be entirely removed.

. But we have now more proof than even this. Reefs of rocks are constantly appearing in places where there were none previously. At Rivoli Bay the soundings have altered to such an extent as to make a new survey requisite. It was known that outside this bay there was a reef of rocks running parallel with the shore, but with sufficiently deep water upon it for small ships to pass over. It is now stated that scarcely any vessel can pass over it, and that some of the rocks have actually appeared above water.

Not very long ago, a schooner, named the *Norah Creina*, was lost upon that part of the coast, and the master of the vessel stated that the rock upon which he struck was not marked in any chart, and though he had been a very long time upon that coast, he had never seen any signs of a reef there before over which a small vessel could not pass in safety.

Again, at Cape Jaffa, to the north of Guichen Bay, there is a dangerous reef, which was marked by the French surveyors more than fifty years ago as extending seven miles from the shore. Some four years ago, a fresh survey was made by the South Australian harbour-master, and the reef was found to exist twelve miles from the shore, and a beacon was erected thereon at that distance. I am now assured, by those well accustomed to this part of the coast, that the reef extends two miles beyond the last distance, and I have seen broken water at least a mile beyond the beacon.

Nor is it alone to this part of the coast that upheaval has been remarked. It would appear that a vast movement is taking place in the whole of the south part of Australia. In Melbourne, the observations of surveyors and engineers have all tended to confirm this remarkable fact; in Western Australia, the same thing is observed; at King George's Sound, the same. As, however, these observations are numerous, I must confine myself alone to the colony to which they refer.

In 1855, a railway was in course of construction between Port Adelaide and the city of Adelaide, between which two places there is a gently-rising plain, about eight miles across. Mr. Babbage, the chief-engineer, who made the surveys for the line, published a paper to show that there was an actual difference of level of some inches between his first and his second survey of the respective heights of Adelaide and the port. As the difference was so small, of course this result cannot be given as certain, because, in eight miles of levelling, errors might easily creep up to that amount.

Under the city of Adelaide there is a thin deposit of shells, containing many recent species, and I have found on hills (many hundred feet above the sea-level beyond Adelaide) a thin deposit of limestone containing shells of recent species. All the hills around are covered for some distance, at least above their base, with limestone; and on Tapley's Hill, about ten miles to the south-east of Adelaide, there is a cutting in the road, about 1,000 feet above the sea-level, which shows a stratum of limestone, about a foot thick, lying unconformably on highly-inclined slates. Though I have met with no fossils in this, I have no doubt that it is of the same period as the limestone on the coast, and shows that the hills have been raised from the sea within a very recent period.

The rivers in this part of South Australia all show very clearly the same fact of upheaval. It has already been stated, that there are not many of these geographical blessings in South Australia, and those that are called so are more deserving of the names of creeks than rivers, with the exception of the Murray. This latter contains most undoubted proofs of the upheaval of the land. When Sturt first sailed down it in 1829, he remarked that the banks must have formerly overflowed to a much greater extent than they did in his time, because on each side of the actual channel there was a flat of marshy land, or else of good soil, bounded on alternate sides by bluff headlands, all of which appeared to have been shaped out by the river, though it did not seem to come near them at the time Sturt passed. These appearances certainly showed that the channel had been narrowed, but not that there was at any time a greater flow of water in the river.

Most of the other Australian rivers which have sufficient water in them to wear much into the soil show the same feature. The Glenelg, which runs a little to the east of the South Australian boundary, has sometimes very large flats on each side of the stream, which has generally pretty steep banks, and from the end of these flats hills rise in some places to about 150 feet above them. It is evident that the water once shaped out not only the flats, but divided the hills which bound them. It is true that the river rises in winter some feet above the flats, or at least it has done so in very wet seasons, but it never comes to the foot of the hills, much less could it have given them their present shape.

Not only here, but also at the Murray river, there is ample evidence that the cliffs on opposite sides of the river were united, and, what is more remark-

P

able, they are sometimes composed of pretty hard limestone, showing very clearly that the river, if it cut through them, must have had a long time for its operations.

Again, on the river Wannon, a tributary of the Glenelg, there are beautiful alluvial flats on each side of the stream, and some parts are studded with round hills, which prove, by horizontal indentations round them, that a stream of water gave them their present form. From all these circumstances, it was very natural to conclude that this country had been subject to greater inundations than it is at present, and this was, in general, the way in which the above appearances were accounted for, but the real cause has been upheaval. It can easily be seen, that when the inclination of a river channel is but slight, the waters will cover a larger area, but with a less depth; but, as the fall becomes greater and the current more rapid, it will have more effect upon the ground, will rapidly scoop out a deep bed for itself, and narrow its channel, which will of course be deeper in proportion to its narrowness.

That something like this does take place may be seen from a river in its early stage of developement, not very far from the coast. At the foot of Mount Graham, about forty-five miles from Guichen Bay, there is a large morass of very deep black mud. This trends away along the east side of a range of hills, in a north-westerly direction, until it becomes, in a mile or so, a perfect channel, about half a mile wide, containing little or no water, but very boggy, and covered with reeds. It continues on in the same width for many miles, until it becomes a stream, which empties itself into the Salt Creek and thence into the Coorong. In winter, a small amount of water drains off in the centre of the morass after the first five or six miles, and the stream becomes more copious as it proceeds farther, but the general character of the creek is a great morass, many miles in length, and varying in width from half a mile to 200 yards, and running for its whole length at the foot of the range. There can be no doubt that as the land becomes more upheaved, and the river has a greater declivity down the coast, the drainage will be better, so that not only will water flow more rapidly, but there will be a larger quantity to run through it. This will not be long scooping out a deep bed in the soft mud which at present lines the bottom, and then there will be presented the same appearance which is assumed now by most of the Australian rivers.

The range, at the foot of which the Creek now lies, will be separated from the stream by a low reedy flat, sloping down to the precipitous banks which will bound the water. The same flat will be present on the other side, each of them probably indented with marks of various water-levels, and then it will seem as if the country were subject to extraordinary inundations, swelling the river to half a mile in width, when, in reality, the appearances are due to there being at one time scarcely any flow of water at all.

When we bear in mind the state of such embryo rivers as the Reedy Creek, we come to understand easily how it is that the banks of some streams are composed of high cliffs of soft earthy clay, which, as they sometimes fall in from floods or other causes, disclose the bones of land animals and freshwater shells. Such remains as these must have become embedded when the stream was in its first stage of formation, when the still water sank deeply into the underlying rock, and decomposed it, mingling its own decomposed vegetable soil with the rocky clay, and giving rise to a morass, in which animals became buried.

This actually takes place in the Reedy Creek at present, for it is not at all uncommon for cattle and horses to become 'bogged,' and to die in the mud, either in an attempt to reach water in the summer weather, or from feeding on treacherous ground. I do not know whether there is any other place where a river can be seen in the very first stage of its formation, and the cursory examination that I have been able to afford convinces me that a great many anomalies in the post-tertiary period might be cleared up by an attentive examination of what takes place during a rapid upheaval of the land.

The mention of rapid upheaval reminds me of a question that might be asked, namely, whether there is simply evidence that upheaval has taken place within a very recent period, or whether it is thought that the process is still going on.

The facts I have mentioned with reference to the appearance of reefs, the alteration of the soundings, the drying up of the lakes, would seem to bear out the view that the process *is* still going on. Add to this, the shocks of earthquakes have not been at all uncommon in various parts of the south coast of Australia, and these phenomena are generally supposed to be more or less connected with actual upheaval.

A severe shock of an earthquake was felt in Melbourne in 1855; another severe shock was felt in Adelaide in June 1856.* Slight shocks have been felt from time to time in various localities to the north of Adelaide; and there are many records of earthquakes having been felt in different parts of the three colonies within the last twenty-two years; so that it would appear that the present is not a tranquil period in the subterranean forces, but that they are still in activity, and upheaval is still going on. It is not to be doubted, however, that there have been many periods of rest since the upheaval first commenced; indeed, there is actual evidence of many such periods, some of which must be noticed as bearing directly upon the country already described.

At the head of Spencer's Gulf, to the north-west of Adelaide, where there are evident signs of up-

---

* A smart shock of an earthquake was felt on the Stone Hut Range in December 1861.

heaval, such as the reduction of the gulf to a very narrow channel of about two miles and more from the high cliffs which bound it, there are also unequivocal signs of long periods of rest. These are shown in three deep indentations which form parallel lines in the cliffs which bound the gulf, and run along it as far as the eye can reach. These may be presumed to have been caused by the water, which ate deeply into the cliffs during a long period of tranquillity.

The evidence of the same tranquil periods occurs in the district to which this book refers, but they are neither so obvious, at first sight, nor quite so certain.

It will be remembered, that in other chapters this district was described as an immense plain, divided every ten miles or so by ridges which ran in a way which seem to follow the coast line with only occasional deviations. The principal of these are six in number, between the coast line and the colony of Victoria, where they cease. I mentioned that the greater part of them are mere ridges of sand, with limestone rock appearing occasionally between; but what is rather remarkable, they undulate and divide into hillocks somewhat on their western or seaward side, while on their eastern they rise rather abruptly from the plain. Wherever limestone is seen on them, or the west side, it has all the marks of coast action, such, for instance, as borings of *Lithodomi*, circular pits lined with lamellar limestone and other similar signs, besides having the limestone much worn and eaten away into caves, contrary, occasionally, to the dip of the strata. There can be little doubt that the western side of each of these ranges has successively been a beach, and possibly they may owe their origin entirely to periods of rest in the upheaval.

Thus there would have been six periods of rest in the upheaval, during which time the sea had time to heap up sand and limestone into dunes, hillocks, and beds, in the way it is at present seen. It must be owned that this is far from being a certain explanation of the origin of these ranges: they may have been ridges underneath the sea just like the Coorong, which is half upraised at present; but the circumstance which makes them very probably the result of coast action, when upheaval was not going on, is, that they seem to follow the coast line, and nowhere rise to a height to which the surf could not have gradually raised them. It is admitted, however, that these reasons are not completely satisfactory, more especially as the width of the ridges and the valley occurring in them would point to upheaval as still going on while they were forming. The circumstance is mentioned, however, just as an observation which has occurred to the writer, which future geologists may confirm or dispute.

It would appear, from some observations that have been made, that during periods of rest the sea encroaches on the land and scoops out the shore in such a manner that they form terraces when upheaved. Nothing of this kind is observed in the ridges, but on lower and more level parts of the coast these terraces are common. Thus, to the south of Rivoli Bay, as far as Cape Northumberland, the coast is very low and flat, only occasionally dotted with rocks, which seldom rise to more than twenty feet above the sea. There is little or no beach, and the waves seem to wash the foot of a terrace raised about fourteen feet. Sometimes the foot of this terrace is a deep bed of black flints encrusted on the outside, in every respect similar to the chalk flints of England. This is the prevailing character of the beach, but here and there the shingles are absent, and a deep bed of loose yellow sand takes its place.

Now, above the beach line, about, as before observed, fourteen feet high, there is a terrace a quarter of a mile or more wide, and as level as a bowling-green; it runs a good way parallel with the coast, but is interrupted on the south by swamps, and on the north by Lake Bonney. This latter, as remarked above, has high sand-hills between it and the sea, but there can be little doubt that it forms part of the terrace now described, as both must have been covered by the ocean about the same time. The terrace seems to have a sloping inclination towards the inland boundary, which is rather an abrupt wall of limestone, about ten feet high, also running parallel with the coast.

The summit of this is another terrace, but it is not so level as the last, and, being rather thickly timbered, its dimensions are not so readily ascertained. It is bounded, however, at about four miles from the sea, by a limestone ridge, which is continuous with the Stone Hut Range, and resembles it in all respects. These terraces are either the result of rests during the periods of elevation, or they may have been sudden upheaval by shocks of earthquakes at a time when Mounts Gambier and Shanck were in eruption.

Though this is not the nearest coast line to them, they are only about twenty-five miles distant. It must be mentioned, that the surface of these terraces is generally stony, not, however, in broken, detached masses, but the limestone lies in flat slabs, much water-worn on the surface, just as if the sea had consolidated the limestone paste and worn it smooth. There are no shells on the surface —at least, I could discover none—which is the more singular, as in the sand-hills on the coast, at a much higher elevation, shells of existing species abound.

Notwithstanding, at the foot of each cliff there is the usual deposit of chalk flints much rounded by attrition, not continuous, but scattered here and there in sufficient quantity to make their identity with those on the coast a matter of certainty.

It will be worth while to enquire for a moment whence these flints have been derived. There are none in the rocks now on the coast, and none, apparently, in those which lie beneath the sea; for the structure of all those which I could examine was quite similar to those described in the last chapter, the Upper Crag, best seen in Guichen Bay. This, we have seen, is composed of small particles of shells and sand, either brought down by an ocean current or deposited on a sandbank. There are immense quantities of flints in the lower crag about Gambier, and those on the coast are in all respects similar. One would imagine, therefore, that the crag only extends a small way from the shore, and the coralline beds crop out in its place, from whence these flints are washed out and thrown upon the beach.

It was much to be regretted that no levels were ever taken from the coast to a certain distance inland. Not being possessed of any appliances of the kind, it was impossible to tell the height of the terraces on the ranges, except by guess-work. It is therefore only a surmise, that the terraces slope up to each other. There was also another surmise, which I only give as a guess, but which seemed to be borne out by one or two circumstances, and that was, that the flats between the ridges sloped inland in an upward direction, and that the flat on the east side was slightly higher than the flat on the west or seaward side. If this were the case, there would be one more argument in favour of the position, that the ridges have been thrown up during periods of comparative rest, during the general upheaval of the land, but it must only be considered as a surmise until a regular series of levels is taken.

In concluding this chapter, the sand dunes of the coast must be mentioned as bearing on a great deal that has been said in the preceding chapter. It has been already frequently stated, that the whole coast from the river Murray to Cape Otway is low and sandy; indeed, this is the prevailing appearance on all the Australian coast. The sand is of three kinds: either in high ridges well grassed, and more or less interspersed with shells; in high detached hills, either bare or covered with salt bush; or in dunes or ridges which are destitute of any vegetation, and therefore liable to drift by the force of the wind in all directions.

Perhaps all the sand now seen on the coast has been originally drifted up, and has only ceased, here and there, by the growth of plants upon it, which has given it firmness and consistency. The dunes, however, are very common, and give a marked character to one portion of the coast, from the mouth of the river Glenelg to Cape Bridgewater; they form immense masses, in many instances three and four miles from the coast, and rising to an altitude of 300 feet, or even more. Nothing can be more dreary than to stand on one of these eminences and gaze below; it is an arid waste of yellow sand, heaped together in ridges or rounded hills, without a patch of vegetation, while afar off the sea rolls on with a heavy surf, making the air resound with its roarings, or terrifying one with the height of its huge crested green waves. On windy days no prospect can be obtained, for then the dunes seem as troubled as the ocean; every gust of wind raises huge clouds of sand, which curl, and break, and drift along, so as to obscure the air.

Valleys are filled up and hillocks swept away, leaving in a few hours scarcely one feature of the former outline.

The rate at which the dunes are encroaching on the land is quite surprising. About half way between Cape Bridgewater and the Glenelg there is a high range, six or seven miles from the sea; between this and the dunes a road runs—the coast road between Mount Gambier and Portland. The sand abuts on the road as a high wall, ranging from 200 to 300 feet high, and the wind brings it down the slope, and of course encroaches more and more upon the space between the coast and the hills. Every month the course of the road has to be altered, and the old tracks serve as landmarks, from which it can be seen that within a few years the dunes have encroached many yards; nothing stops their course. Bushes are covered in a very short time, large trees are surrounded and buried before their leaves have time to wither, and here and there, what appears a bundle of twigs sprouting out from the sand, is nothing but the top of a high gum-tree which had been heaped over, and all but this 'in memoriam' is covered. The sand, when examined closely, is found to consist of very small fragments of shells, too minute to allow the least chance of identification, and clear grains of silicious sand.

In no place that I was able to examine could I find the smallest indication that the sand became consolidated into a rock, or of any concretions formed by the percolation of rain or surface water.

The only sign, indeed, that rain had made any impression, was at the edges of the slopes, where it cut trifling little courses, and caused the sand, here and there, to slip. These facts are the more important, as they afford additional reasons against the subaerial origin of our Upper Crag. If this rock had been formed out of water, its thickness and general character would indicate something very similar to the sand dunes, and then the concretions which are met with must have been caused by the infiltration of rain water. Now, for rain water to have formed concretions more in one place than another, it must have collected on the surface, but this it would not do in sand like these dunes, which absorbs water equally on all parts of its surface. True, if the top was covered with trees it might have collected, but there are no trees on such accumulations of sand.

Some, however, may think that the burial of trees may be the origin of the concretions in the crag, and offer another reason in favour of its subaerial origin; but to this it may be replied, that in the absence of any instance here of the hardening of these dunes into a rock, such a theory is not consistent with observed facts, though such may be the case at Bald Head and Cornwall.

The strata observed by Mr. Darwin may have been hardened trees, but I could not here find any traces of the same process.

Again, the quaquaversal dip of these strata, the oblique lamination of the crag, has never been seen by me where a section of the sand dunes was exposed; on the contrary, nothing but a homogeneous mass of sand was perceptible. I am far from denying, however, that if the dunes were hardened into rock, and the trees and branches transformed into calcareous casts, the appearance of a section would, in some respects, bear a strong resemblance to the Upper Crag: but the marine origin of the latter is strongly evidenced (as shown in the preceding chapter), either by its always existing on a coast which has been quite recently upraised from the sea, or by its being in many instances covered with marine shells in limestone, or with trap rock which flowed under the sea, and a subaerial course had nothing to do with its formation.

It is rather singular, however, that wherever sand dunes are found, a calcareous sandstone formation, like the crag, is generally noticed of the coast below it. This is the case on the east coast of England, as also near Lake Superior, in America, where immense sand dunes in the coast are bounded by a hardened calcareous sandstone rock. Again, in South America, Darwin, in mentioning the sand dunes of Bahia Blanca, mentions as near them the great sandstone plateau of the Rio Nigro: the latter, from the great distance of the two localities, is hardly a case in point.

A moment's consideration will show why, perhaps, these phenomena are always associated together; not, however, because the rock is derived from the dunes, but because the latter are derived from the rock. Thus, sand dunes are found near old red sandstone, near secondary, and near recent calcareous sandstones, but in every case it is the weathering and decomposition of the rock whence the sand is derived, and this is the reason why it is found in such large quantities as to drift into hills, valleys, and ranges.

To my mind, it would be just as absurd to say that the Old Red Sandstone has been formed by the hardening of the modern dunes, as it is to say the same thing of our Upper Crag.

In conclusion, some apology must be offered for having dwelt so long on a point of apparently minor importance; but when it is remembered that this formation is found on a great many parts of the Australian coast, nearly, in fact, encircling Australia as a belt, it becomes important to settle the question of its origin.

Possessing some connection with the coralline strata underneath, and lying on the coast with the most evident marks of upheaval on its surface, it belongs especially to the subject I have attempted to describe. It is the last and uppermost of the stratified series; and, having dwelt on it and on the subject of how these rocks ever came to be displayed to us from beneath the sea, it remains to consider the volcanic evidences of the district, which will be the subject of the next chapter.

CHAPTER VIII.

EXTINCT VOLCANOES.

PRELIMINARY REMARKS.—ABSENCE OF VOLCANOES FROM AUSTRALIA.—PROBABILITY OF LESS DISTURBANCE IN SOUTHERN HEMISPHERE.—MOUNT GAMBIER.—BY WHOM DESCRIBED.—THE LAKES.—THEIR PECULIARITIES.—THE VALLEY LAKE.—THE PUNCH-BOWL.—THE MIDDLE LAKE.—THE BLUE LAKE.—MODE OF THE VARIOUS ERUPTIONS.—VOLCANO ONE OF SUBSIDENCE, NOT UPHEAVAL.—MINERALS FOUND IN THE CRATERS.—PERIOD OF THE ERUPTION.—PROBABILITY OF ITS EXTINCTION.—RECAPITULATION.

WITH the last chapter we have concluded the natural history of the sedimentary rocks of the district, and we therefore pass to others of a different origin. It will be necessary again to notice circumstances and phenomena connected with the aqueous formations, such as caves, deposits of bones, &c., but as these are the result of changes not connected with the origin of the rocks, and to which both igneous and aqueous deposits may have been equally subject, the description of them will be more proper at the end of this volume.

Let us therefore now turn to the igneous rocks of the district.

It has sometimes been remarked, that Australia, for its size, is possessed of fewer volcanic remains than any other country of equal extent, while

Europe, a continent not very much larger, contains several, which are even now in a state of activity, and is literally studded all over with extinct craters. Australia, as far as it is at present known, contains none of the former and comparatively but few of the latter: probably a reason will be found for this when the geology of this continent is more studied. At present, I feel convinced that it is one of the many evidences we have that disturbance has been much more frequent in the northern than in the southern hemisphere. Look, for instance, at the immense extent of the formations in South America—meaning, of course, the fossiliferous formations. There is the great Patagonian tertiary formation, extending (according to Darwin and M. d'Orbigny) from St. Cruz to near the Rio Colorado, a distance of 600 miles, and reappearing over a wide area in Entre Rios and Banda Oriental, making a total distance of 1,100 miles; and even this formation undoubtedly extends south of St. Cruz, and, according to M. d'Orbigny, 120 miles north of Santa Fé. In addition to this wide area, there is the Pampean formation, celebrated as the sepulchre of the bones of the mastodon, glyptodon, megatherium, &c., which extends over many degrees of latitude. In our own continent (Australia) we have formations nearly as large; — witness the coralline strata described in the previous chapters (the Crags).

Now, such immense and uninterrupted formations are not known in Europe: on the contrary, the amount of different deposits to be found within a small area is surprising; and these are broken by faults, dykes, and inclinations, showing great disturbance, even where the strata are continuous. The best proof that could be given of the greater disturbance in the northern than the southern hemisphere, is that in Europe — nay, perhaps, almost in Great Britain alone — all the deposit of any geological epoch may be studied; but, supposing that geology had just been cultivated in Australia, the whole secondary period, from the New Red Sandstone to the Chalk inclusive, would have been left out of the classification, because such deposits are quite unknown there, and repose and tranquillity would rather be supposed to be the rule of Nature's operations, than the 'immense catastrophes' which earlier geologists were led to infer from what they saw in Europe.

I feel convinced, therefore, that further investigations will show that disturbance was uncommon in the southern hemisphere, in comparison with the northern; and this fact, when established, may lead to revelations of subterranean agents, the importance of which we cannot foresee. In the mean time, we must content ourselves with close observation and a record of facts, feeling certain that theory and generalisation will easily be accomplished when the hard work of detail has been got over.

With such a view, we have now to record observations on the igneous rocks of this district, commencing in this chapter with the remarkable extinct crater of Mount Gambier; and probably the particulars are interesting to science, not only on account of its being one of the most extensive in South Australia, but because a faithful description of it may serve as a guide to other volcanic phenomena on this continent.

The ground has not been previously quite untrodden. Captain Sturt, as I am informed, made a series of observations on the place, but did not, as I am aware, proceed any further with regard to publishing his remarks. It is supposed, however, that he made some communication to Mr. G. P. R. James, and, accordingly, a rather romantic and incorrect account of Mount Gambier has found its way into one of the works of that novelist.

In 1851, Mr. Blandowski surveyed and mapped the three lakes, and made some valuable observations on their mineralogical and geological peculiarities. Part of the latter were embodied in a series of letters to the Adelaide German Newspaper, but, owing to the gold discovery, and the confusion subsequent thereon, the maps, &c., were, I believe, unfortunately lost. Nothing further has been done in the exploration of the crater. I have not seen Mr. Blandowski's papers on this subject, and therefore cannot say how far his views and mine coincide; but, should any of my conclusions bear a stamp of less probability than any he has advanced, I shall be most happy to give way as far as possible, as the object of this work is the advancement of science, and not my opinions. Everyone caring for truth will of course always prefer a true theory to giving currency to any deductions of their own. However, I am sure of this, that the facts are strictly stated, as observed; and as I have always given the reasons which have led me to draw any conclusions, readers can judge for themselves whether they are hasty or not. With these prefatory remarks, let me proceed at once to my observations.

The extinct volcano, which is included in the general title of Mount Gambier, is a chain of craters extending nearly, but not quite, east and west; the wall on the west side being by much the most elevated.

There are three lakes, and they possess such distinct features that they require to be described separately: that on the east end, called the Blue Lake, is a large and deep body of water of irregular oval shape, whose longest diameter is nearly east and west. It is surrounded on all sides by banks between 200 and 300 feet high, and these so steep and rugged that descent to the water's edge is quite impossible, except in one or two places. The sides are thickly wooded with varieties of the *Melaleuca* (the tea-tree of the colonists), excepting where the rough rocks stand out in perpendicular escarpments, and thus the dark-green brushwood is broken by huge and craggy rocks descending precipitously for forty or fifty feet. These crags sometimes hang over

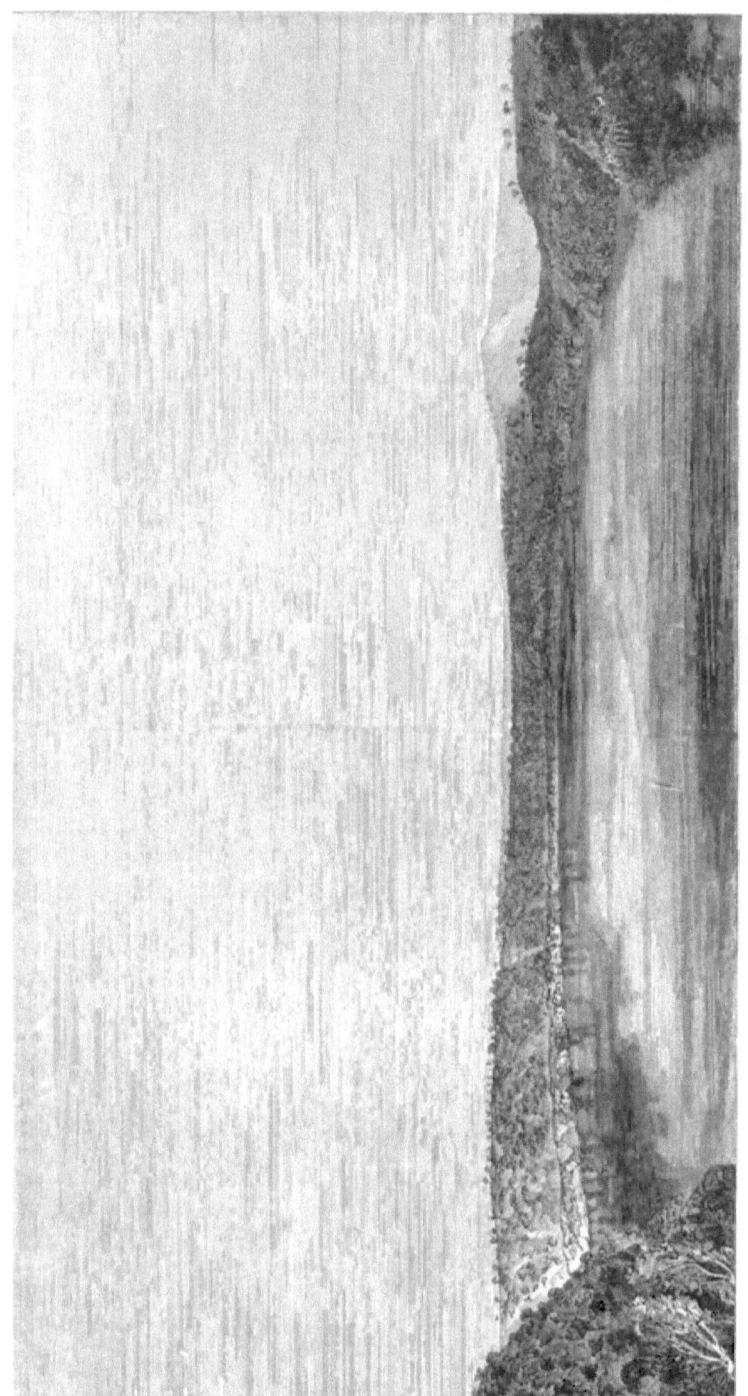

MOUNT GAMBIER. BLUE LAKE CRATER.

the water, whose already dark-blue tint is rendered still more gloomy by the reflection of their black and stony fronts. The whole appearance of the lake is wild and sombre in the extreme. The deep-blue, or rather inky appearance of the waters, the blackened precipices which bear so plainly the tokens of fiery ravages, the thick and tangled nature of the brushwood, give the place an air of savage loneliness; and then the place is so quiet, so still, that, but for the cawing of the rooks overhead, or the splashing of a solitary water-fowl, one might almost imagine Nature to be at rest, tired with sending forth those volcanic fires which poured forth ages ago.

Looking at the walls from any side, four distinct kinds of rock are visible. There is, first, the larger ash, decomposed into soft black surface soil, covered with grass and trees, and varying in thickness from forty to seventy feet. It extends, in some places, to the water's edge uninterruptedly; underneath this there is a precipitous escarpment of black lava, generally forty feet thick, but at the western end of the lake much thicker and more precipitous; this extends nearly all round the lake, and is very seldom inclined or broken, or in any other way than a precipice, rough and jagged, and having no dip towards the water. Under this there is, in places, a layer of greyish-brown ash, about two feet thick and very finely laminated: this is only occasionally seen. Beneath this there are about twenty feet of coralline rock, full of fossils, and belonging to the Mount

Gambier Lower Crag formation, with the strata quite horizontal, and bearing some marks of having been exposed to a high temperature, but rarely crystallised. This latter bed of rock forms a well-defined white line, nearly continuous round the lake, at a uniform height of, perhaps, rather more than twenty feet. The lava is not vesicular, or rarely so, and seems to have flowed from some of the lakes to the westward, about which more will be said presently.

The next lake is merely a good-sized pond, of moderate depth. The level of the water seems about the same as the last lake, and the banks as high, if not higher. They are not precipitous, but slope all round to the water at an equal inclination, with little or no outcropping of rock. They are well grassed and studded with shea-oak (*Casuarina æquæfolia*) and honeysuckle (*Banksia integrifolia*); the water at the bottom has only made its appearance, as I am told, within the last few years. It must not be forgotten, in reading the description of these lakes, that they are joined together, so that the west walls of the Blue Lake make the eastern ones of the Middle Lake, as it is called. There is a break, or rather a deep indentation, in the height of the walls between the Blue and Centre Lakes, so that a person standing on the centre of the partition between them sees the walls on his right and left slope upwards from him: this is seen in the foreground of the engraving as a kind of pass leading from one lake to the other. The same thing occurs between the Centre Lake and most westerly or Valley

MOUNT GAMBIER. MIDDLE AND VALLEY LAKE CRATERS.

Lake; the walls, then, of the Centre Lake are highest on the north and south, while on the east and west line, where they join the two others, they form deep depressions or passes between, though still at a considerable height above the crater. The height of the lowest part above the water is probably, from a rough calculation, about 170 feet, and the highest perhaps double that.

The third lake differs much from the other two, and is possessed of so many and such varied features, that it becomes difficult, in the details of these, to give a good general idea of its appearance; it is larger than the Blue Lake crater, and of almost circular form, but the bottom is only partially covered with water, very deep at the east end, but shallow on the west. Those parts which are left dry are always connected with the sides (which are lower there), though in one instance by a mere strip of land, and the ground is very undulatory, rising, at times, into hillocks, which are some little height above the water. The water is at each end, and the ground in the middle, but by far the largest lake is on the eastern side. In the dry part, there are three ponds, which, being circular, appear at a distance like wells sunk side by side. The view from above them would incline one to call the Valley Lake crater a basin with strips of land, which are covered with little ponds, and have a very uneven surface.

The crater walls surrounding this lake are very remarkable. At the eastern end they are lowest, rising gradually till about a third of the way round on the northern side, and then, rising suddenly into a peak and descending again for a short distance, again mount, by a very abrupt elevation, to nearly double the previous height, from which point there is a gentle slope upwards to the highest part of the mount, where a trigonometrical station is erected. From this there is a still more abrupt descent to the usual height of the sides, which is continued round to the starting-point at the east side. That part of the wall which is so considerably higher than the rest is what is properly termed Mount Gambier. (This is the peak seen in the engraving.) It is the higher wall of the crater, and gives a better key to the kind of eruption that has taken place than any other part of the mount.

Standing on the highest point, one perceives a basin on the south side which is called the Punchbowl. It seems like a hollow scooped out by an eruption in the side, and at a distance appears precisely similar to the Cumbrecito in the side of the Caldera, in Palma (Cape Verd Islands). On nearing it, it is found to be very deep, so that its real form is like a funnel, with one side (that which is inside the lake walls) much lower than the other. Here a sort of pitch stone porphyry is very common, especially on the lower or inner side.

At first sight, this appeared to be a crater on a small scale, and such no doubt it is, but there is no sign of any tufa around, as if having fallen from a centre, and the soil is so deep on the inside and so covered with long grass and fern, that assertions as to its origin are founded alone on its shape. The occurrence of such little craters, either at the side or in the walls of craters, gives rise to many speculations. It does seem strange, that while a central large crater is carrying off the subterranean fires, any other vent should be formed so close by. Possibly it may be one of the many cracks formed at the same time that the crater in the centre was opened, and the steam and gases issuing therefrom would prevent any deposition of ashes upon it while they were deposited all around.

Or it may have been a small crater established subsequently. These are very common, even when there is a large central point of ejection. Thus, Mount Etna is surrounded with small cones; Vesuvius, Ischia, and Hecla are also innumerable instances of the kind. It must be said, however, that, had it been a subsequent eruption, more disturbance in the original walls would be seen. It may be remarked, in passing, that earthquakes alone have been known to produce such funnel-shaped hollows. The small circular ponds in the Plain of Rosarno, caused by the Calabrian earthquake of 1783, are cases in point.

A little past this Punch-bowl, as it is called, nearing the eastern side, the walls, instead of sloping down to the lake as heretofore, become precipitous, and the volcanic ashes, disappearing from the sides, appear only on the top. The precipitous portions of this end (forming at least half the wall) are formed of the fossiliferous coralline strata peculiar to this district. The stratification is nearly horizontal, with a decided though slight dip inwards towards the lake; which dip, being quaquaversal, or inclining all round to a common centre, shows the rock to have subsided into a hollow previous to the erupti n, of which hollow, as will be afterwards seen, the present crater forms only a part. The water appeared to me to be deepest here. The strata are not in the least altered by fire, as far as one could judge from a short distance, the only change being a weather-worn appearance, which is observable in caves elsewhere.

There is no channel or dyke in any one of the three lakes such as would have been made by a flow of lava; indeed, there is very little appearance of lava in the whole group, with the exception of the stratum, which here, as at the Blue Lake, lies between the ash and the coralline strata. But it is only a moderate seam at the end of the Valley Lake just described. The lava where it is found varies in its character, but may be described as dolerite, sometimes very porous and scoriform, in which case it is of a bright brick-red colour; generally, however, it is a blackish brown, occasionally enclosing crystals of glassy felspar and augite. Fragments of scoriæ are found on the sides and bottoms of all the lakes, with pieces of lapilli and porphyry (black base and glassy felspar). On the sides near the higher wall or mount, scoriæ occur more frequently, evidently having streamed down in a backward flow, before cooling. The fracture of these pieces is smooth and glassy, exactly like pitch or new coal and coke.

To return now to a more minute description of those portions which throw a light on the past history of the volcano, we come to what is termed properly Mount Gambier, and is, in fact, the highest wall of the Valley Lake crater. This is formed of successive layers of an ash conglomerate, composed of scoriæ, fragments of obsidian, porous lava, and pieces of the fossiliferous rock, all cemented together into a very hard stone. The wall is a mere ridge on the top, but slopes down on each side to a considerable thickness. The rock not being decomposed, the layers are well defined. A good section is seen of the highest wall or cone from the inside of the Valley Lake. This is rendered still more conspicuous from the occurrence of a ridge, or sort of buttress, which runs from the water edge to the very highest point of the summit. It is here observed that the layers of ash or strata thin out rapidly, and are inclined at a greater angle in proportion as they near the summit. Thus the dip is constantly variable. This gives a satisfactory answer to the application of Von Buch's theory in the case of this volcano. Other and more decisive reasons why this cannot be a crater of elevation will be subsequently given. What is somewhat remarkable is, that the strike of these strata inclined at each side of the buttress already alluded to. This ridge, as it may be called, comes out pretty considerably into the lake at its foot, and it is matched on the other side by a similar promontory, making the ground plan of the lake like the figure 8. This fact is of importance, because as these two promontories make nearly a complete circle round the water where the volcano has left most traces, it would seem as if the eruption was confined to that end only, at least latterly in its history. This will appear more reasonable from what will be hereafter advanced. The activity of this part of the crater must have concluded with as great violence as the commencement, because at the very highest part of the ash cone there are two or three immense fragments of the fossiliferous rock embedded firmly in the conglomerate.

It has been already observed, in describing the walls of the Valley Lake crater, that the higher wall or proper mount rises abruptly above the ordinary level of the walls. Before this takes place, there is on the north side an isolated hill or hummock forming part of the wall. Between this and the higher wall the sides are not precipitous, but slope down into a kind of terrace or half-basin, which near the lake becomes a small precipice, covered with red scoriæ. At the foot of this small escarpment the water is not reached, but there is a gently-undulating ashy slope down to it. This half basin, with its isolated ash cone, forms somewhat of an inlet from the general form of the lake. Particular attention must be directed to it. Evidently, it has been a crater; probably, oldest and first of all.

From this crater it would appear the lava has been derived which lies about the limestone in the Valley and Blue Lakes; for from this point the walls on the north side are higher and more undulatory, and, wherever sections can be seen, are formed principally of thick black lava, flowing away in the direction of the Blue Lake. Considering, however, the moderate thickness of the stream, and its being so slightly vesicular, it must have flowed in a highly liquefied state, spreading out into a sheet, owing to the level character of the ground, and covering spots now occupied by parts of the Blue and Valley Lakes.

Thus far I have been merely describing those features which are calculated to elucidate the geology of the volcano; I must now consider, from these evidences, what kind of eruption has taken place to cause the appearances related. Let us take the Valley Lake first, as best fitted, from its peculiarities, to give us an insight into the whole phenomena. In the first place, we have already seen that the upper part, or west end, contained a recent crater near the ridge, and the relics of an ancient one on the north side.

The east end of the Valley Lake, it has already been remarked, has precipitous walls of limestone, lava, and ash, with deep water at the foot. The bottom of this water is ashes and scoriæ: I do not think, however, that the whole Valley Lake has been a large crater at any time. The eruption of the crater was, in my opinion, entirely confined to the west side, and was neither, comparatively speaking, very violent, nor of long duration. The greatest height of ash is probably not more than 600 feet, and this appears to have been formed almost exclusively from the deep and irregularly-formed lake which lies at the bottom of the higher wall, and this wall, moreover, is nearly the only remains which the eruption has left behind to mark its progress. Of course the two little well-shaped vents for the middle of the lake, already alluded to, contributed their quota; but, as they are both surrounded by a circular wall, some ten or twelve feet high, one must regard their contributions as not on a very extensive scale. The east end, where deep water washes the precipitous banks, may have thrown out some of the ashes that are found on the banks above it; but the origin of this portion of the lake will be considered by and by.

That the eruption of this crater was not very violent, may be gathered from the following facts. The tufa, &c., are not scattered very far, and do not seem to have been thrown to any considerable height; for the higher wall is so near the point of ejection, and so very narrow, compared with the sides of the Blue Lake. Again, the strata of conglomerate thin out so rapidly, that they could not have been formed by a volcanic process on a very large scale; besides, to sum up all to a self-evident proposition, if there was much thrown up there would be more to be seen than there is at present. Many who have seen the mount will be surprised at its size being considered small, but when we remember that the volcano of Jorullo, in Mexico, was elevated considerably over 1,000 feet in a single night — when we remember the tremendous height of some volcanic mountains, Mount Loa, for instance, by no means the largest, which is 4,000 feet high — we cannot think it would take very long to form a cone of the moderate pretensions of Mount Gambier. Of course, when we speak of small eruptions and moderate, these remarks must be qualified by recollecting that any volcanic action is the result of a vast convulsion of nature, always attended with serious effects; and, had there been any life or property in the neighbourhood of Mount Gambier at the time of its eruption, the results would doubtless have been quite extensive enough for the sufferers. But it may be asked, how is it, if the disturbance was confined to one side of the crater, that the ejectamenta do not form a complete circle round the point of ejection? The answer to this difficulty, which required some investigation to solve, explains one of the peculiar features of the volcano, which is, I think, unparalleled in any other volcano in the world, and accounts for the fossiliferous precipice on the eastern end of the basin. The hypothesis, which, after considerable enquiry, I have been induced to adopt, is this.

Previously to the second eruption of the ancient crater, but after the first, a large circular mass of limestone fell in, owing to a subsidence underneath; this subsidence was, of course, connected with igneous agents, and as the same phenomenon has occurred at the Blue Lake, we shall consider it more at length when describing that crater. Such a chasm thus formed would be an ample receptacle for all the ejectamenta which fell eastward. This theory received every support from what is observed elsewhere; indeed, it would never have suggested itself had I not observed the phenomenon in other parts of the district. Thus, at a spot about a mile from the Blue Lake, there is a place called the Cave Station (previously alluded to), at which two immense basins of chasms may be seen, whose precipitous sides and many other evidences easily recognised, show them to result from immense masses of rock having fallen in.

The friable nature of the coralline rock renders it much more liable to this kind of accident; and the country, to some distance round, is filled with caves and funnel-shaped holes, which nearly all owe their origin to the same cause. That there was a subsidence at the Valley Lake after the upheaval of the strata is shown by the quaquaversal dip of the beds towards the centres of the basin, and that a chasm was eventually the result of such a subsidence, is recognised from an identity of appearance with the caves, as they are called, just alluded to.

Probably there was an eruption of ash from the chasm when it was formed, and this explains why there is no line of division or separating wall between the east and west ends of the Valley Lake, for they evidently form separate craters.

The great disproportion of what I have termed the higher wall to the rest of the ash deposit must of course be attributed to the prevailing direction of the wind, which is always very violent during volcanic disturbance; for the air, heated by the boiling liquid below, rises rapidly, and cool air, rushing in to supply the vacuum so caused, gives rise to a current of air in one direction. This is the reason why, in the volcanic island of St. Paul (38° 44′ S., 77° 37′ E.), the west side is 800 feet high, while the east is not much above the water's edge. But as the wind only accounts for a disproportion, and not for the total absence, of one side, the theory of the chasm — which is supposing what the appearance really bears out — is the only satisfactory explanation.

There is one thing more to be added just now: two promontories were spoken of which jut out from the walls and partially enclose the water. One of these promontories is the ridge already described, which runs to the top of the highest part of the mount; the other is very remarkable. Seen from the east, it appears like a succession of nearly horizontal layers of ash, rising into a straight thin wall, nearly forty feet high, but seen from the peak it is found to be composed of strata of tufa dipping in towards the central point of ejection at an angle of nearly 60°. The fact of its having an inclination only one way, and that towards the west or highest part of the crater, is pretty conclusive proof that at the time of its formation there was only one end of the lake from which ejectamenta were coming, and that was the western end. Some subsidence has taken place since the deposition of the ashes found on the north side of the basin, where the amygdaloidal lava is in greatest quantities: the ground sounds very hollow on percussion for some distance, showing the existence of some cave underneath, the hard nature of the pitch stone just there preventing its falling in.

We must now turn our attention to the Central Lake. Whatever has been said of the Punch-bowl, on the south-west side of the Valley Lake, applies equally to this. It is larger, but a mere sloping chasm of half-decomposed ash, with a pond of water at the bottom. It has been stated before, that no rock is visible on its sides. It is well grassed.

The eruption from this crater has not been very violent; probably, it was subsequent to the most ancient crater in the Valley Lake. There are no data to form an opinion as to what relation, in point of time, it bears to the other craters. A section of some little depth on the top of the sides near either of the other lakes would show, by the stratification of the ash, which was prior to the other; and, unfortunately, no such section is obtainable at present. There is nothing, however, against the theory that they may have been synchronous. It is rather strange, however, that this crater lies in the straight line between the other lakes, each of which has a seam of basalt underneath the ash, and there are no signs of this seam in the crater under consideration, neither does the subjacent limestone show. This may arise from the more moderate depth of this crater. I cannot help thinking, however, that it goes deep enough to show both. The absence of the seams may be explained otherwise. If the seam of basalt had flowed prior to the eruption of this part of the volcano, its subsequent breaking forth might have blown away the seam of trap and covered the fragments with ash; and there are fragments in the sides to bear out this hypothesis.

Another theory which has suggested itself is this—the crater may have been in activity while the lava was flowing, and so have heaped up sufficient ash to have kept the stream away from its mouth. In effect, the ash is higher on the side past which the lava flowed. These are the only facts worth mentioning in connection with this crater, which seems to have been quite undisturbed, and remains now like a blackened cauldron, a sombre monument of the ravages of its former igneous tenant.

The consideration of the Blue Lake has been reserved till the last, as being the most extensive, and as where the eruption both began and ended. Having already described its aspect and appearance, we have only now to do with its geological features. From the regularity in the form of the walls, and from their uniform height all round, one easily concludes, that whatever eruption took place from this crater, it was sustained from a line in the centre, without being subject to any variation. Indeed, the whole seems to have been formed by successive layers of ash regularly distributed all round on the top of the stratum of trap (much thicker here) which lies on the limestone, and through which the volcano has broken a passage.

Close to the lake the ash is probably 150 feet thick; at a quarter of a mile this is reduced to between forty and fifty feet, and at the distance of a mile this thins out to a mere seam, varying from three to six feet in thickness, and so on till it becomes lost in the upper soil. This is what is perceptible about a foot or so from the surface, but, as the dark soil of the country is nothing but the result of decomposed ash, the deposit must have been much thicker than it now appears. Nearly all round the lake there is a regular line of demarcation, made by the thick seam of basalt which intervenes between the rock and the ash.

It has been already stated, that a layer of highly laminated grey ash lies between the basalt and the limestone; it is about two feet thick in some places, and the laminations dip in all directions. This clearly shows that an eruption had taken place before the flowing of the lava, since it is underneath it. This could not have been from the Blue Lake, because, in that case, the lava would show some signs of having flowed over into the basin. But it does not. On the contrary, it appears in clean escapements, as if broken away round the edge of the lake after cooling. The general dip of the ash laminations points to the Valley Lake, probably the ancient crater, as the point whence they proceeded. Had the crater, at the point whence they proceeded, been nearer, we might expect the ash deposit to be thicker than it is found.

The limestone underneath was not altered or crystalline wherever examined. This is not surprising. If volcanic sand acts as a non-conductor of heat to such an extent that clefts in Mount Etna, filled with snow and ice, when covered with it are not melted by subsequent flows of lava, we can easily understand why the limestone should remain unaltered. A layer of ash, two feet thick, would amply resist the heat of a much thicker flow of lava than that found at Mount Gambier. I may just mention, in passing, Naysmith's experiment, quoted by Sir Charles Lyell in his 'Principles of Geology:'—'A cauldron of iron one inch thick, lined with sand and clay five-eighths of an inch thick, was able to contain eight tons of melted iron at a white heat. Twenty minutes after the pouring in of the iron, the hand could be placed on the outside without inconvenience.'

The limestone is not altered; the edges of the strata exposed to the lake are discoloured, just as if gunpowder had been exploded, here and there, in spots; the strata are hardened, and detached fragments ring on percussion; the edges are also jagged and precipitous, like the lava above them. There are no incrustations of lava; no pumice or scoriæ adhering to the sides, wherever I could examine them; and, as far as appearances go, this holds good all round. I must observe, however, that, as some portions of the wall rise straight from the water's edge, they can only be examined by means of a boat.

Descending to the margin of the lake, (a proceeding which requires some little nerve and prudence to accomplish), the appearance of the water is quite changed. Instead of having that dark and murky hue it seems to possess as seen from above, it assumes a beautiful crystal clearness, unequalled by the purest spring that ever flowed from a rock. Rapidly deepening from the side, the water becomes a delicate azure at a short distance from the brink, still, in its faint distinctness, showing the outlines of great boulders of rock on the bottom, whose great proportions are gradually lost in the increasing depth. And there the surface is so calm and quiet, only disturbed by the most gentle rippling, which wreaths the pretty water-plants into most graceful forms, and makes them, from time to time, reveal the surface of the snow-white rock upon which they grow.

Sometimes, however, the water deepens almost perpendicularly from the sides. It is somewhat singular, that though the sides are formed either of the coralline rock *in situ*, or of large fragments of this limestone lying on the rapidly sloping sides, there are no fragments of the basaltic trap which lies above it. If, as hereafter will be proved, this crater began its career by the falling in of the chasm now visible, it is strange that fragments of trap are not as common on the sides beneath the water-level as fragments of limestone, since both equally formed portions of the superincumbent mass. Perhaps no fragments of either remained on the sides at the time of the formation of the chasm, and those that are now seen have been detached subsequently from portions which lay under water.

The lake is known to be 240 feet deep in the middle, and from soundings it would appear that the bottom is flat and equal, like a floor. This was ascertained from a boat which took the Governor (Sir R. G. M'Donnell) upon its surface—the only time its waters were traversed by man. I was unable to find any tufa, scoriæ, or porous lava, a few fragments of pitchstone being the only volcanic evidences which appear. This may not be the case all round; but, until more facilities are afforded for investigation, my conclusions must rest only upon what I am able to observe. Here, then, the evidence shows there has been an eruption which has been considerable, both from the size of the lake and the immense quantities of ash thrown to such a distance. That it has been accompanied with violent explosions is seen from the immense masses of basalt which are sometimes buried in the ash, and yet, with all these marks of disturbance, there are no signs of any outpouring of lava, little or no pumice or scoriæ, and not even an aperture in the side through which any lava could have flowed, nor any fragmentary slags adhering to the face of the precipitous rock. Such appearances, seemingly contradictory and inexplicable at first, are consequences of the peculiar nature of the eruption which took place. I am going now to give a history of the igneous activity of the volcano, which will clear up whatever obscurity there appears to rest on the mode of its disturbance, and, as I proceed along, I shall give the complete chain of evidence by which the explanation is supported; but as the theory would appear startling unless some parallel case were cited, let me, by way of preface, give an account of a volcano at present in activity, which Mount Gambier most resembles.

In the Sandwich Islands there is a volcano called Mount Goa, which, though very much larger than the one under consideration, resembles it in many ways. In the side there is a lateral crater, at present in activity, called Kilauea, which is 3,970 feet above the level of the sea, or about the same height as Vesuvius. Sir C. Lyell, in his admirable 'Manual of Geology,' describes it thus:—' Kilauea is an immense chasm, 1,000 feet deep, and in its outer circuit no less than from two to three miles in diameter. Lava is usually seen to boil up from the bottom in a lake, the level of which alters continually, for the liquid rises or falls several hundred feet, according to the active or quiescent state of the volcano; but, instead of overflowing the rim of the crater, as commonly happens in other vents, the column of melted rock forces a passage into subterranean galleries or rents leading towards the sea.'

A Mr. Coan has described an eruption which took place in 1840, when the lava had risen high in the crater and began to escape from it. The direction of the current was first traced from the emission of a bright vivid light from an ancient crater 400 feet deep, about six miles to the eastward of Kilauea. The next indication was about four miles farther on, where the fiery flood broke out and spread itself over about fifty acres of land, finding its way underground for several miles farther, to reappear at the bottom of another ancient crater, which it partly filled up. The course of the fluid then became invisible for several miles, until it broke out, for the last time, twenty-seven miles from Kilauea, running in the open air for twelve miles, and then escaping into the sea over a cliff fifty feet high in a cataract of liquid fire which lasted for three weeks. The termination was about forty miles from Kilauea.

Now, there can be very little doubt that something similar to this has happened at Mount Gambier, consequent on the eruption, perhaps, of both craters. The mount is scarcely fifteen miles from the sea, and being not much above the level of the latter, would not give occasion to the lava to come to the surface during its passage.

Mount Shanck, another extinct volcano, lies in a straight line between the sea and Mount Gambier; but, as it will form the subject of the next chapter, I will not enter further into its description than to state that there is no mark of any lava stream from Mount Gambier in its vicinity,—nor need we expect it, since the igneous forces which caused both must have had a subterranean connection.

The theory that the lava flowed underground into the sea, was formed after investigating the features of the lakes, for it seemed quite natural to conclude, that after such an eruption there must have been a flow of lava in some direction; and I thought it likely, provided the sea level had not much altered since the eruption, there should be some signs of volcanic rocks on the sea coast to the south of the craters. This is, in fact, the case. A seam of trap is seen on some of the rocks, as though it had come to the surface and flowed over them. The trap is not vesicular, and may have flowed under the sea, because this part of the coast has only recently been upheaved. It is not certain, however, that it does come from these craters, though the probability is greatly in favour of that theory.

We will now consider the peculiar features of the Blue Lake as indicating the kind of eruption that has taken place. From the fact of the seam of lava bearing most positive evidence of having been fractured all round, to give rise to the present crater, there can be no doubt that the chasm owes its origin to subsidence and the falling in, *en masse*, of the superincumbent strata. This seems a bold theory, but no other will coincide with the appearances the lake presents. Had the chasm been already there when the lava flowed, it must have shown some signs of flowing over the banks, but none such exist. The rock appears to have been split into a jagged precipice by the falling in of a part. From the crater thus formed ashes and scoriæ were ejected. Its depth by the present soundings, from the top of the lava to the fused mass, must have been nearly 500 feet; of course nearly all the subsided rock would be rapidly fused, except some few fragments thrown into the air by explosions and deposited on the sides. Such fragments, some, perhaps, weighing as much as a ton, are seen embedded in the ash. It must not be imagined that there is any novelty in supposing extensive subsidence during volcanic eruptions. Indeed, the Val del Bove, on the side of Etna, is supposed to have been caused by a similar agency. Mr. Charles Darwin, in his interesting volume on volcanic islands, has given many instances of subsidence coincident with volcanic disturbance, or immediately following them.

It will be remembered, also, that subsidences are supposed to have operated at the east end of the Valley Lake. It must be admitted that there is a novelty in assuming a crater with such an origin, but no doubt can be entertained that after the subsidence an ejection of ashes took place. This mode of eruption of a volcano is hardly in accordance with received theories as to the manner of their breaking forth, nor should I venture to propose it, were it not strictly in accordance with observed facts. Nothing, however, can more perfectly contradict the crater elevation theory as applied in this case. So far from there being any marks of elevation, the limestone strata preserve a most perfect horizontality at the water's edge. Abruptly as the ash dips outwards, the limestone strata most convincingly show that it (the ash) has not been upheaved to its present position.

Instead, then, of an elevation theory, we must adopt a subsidence theory in this case. Whether this is applicable in any other instance I am unaware, but the fact, if new, may be useful in explaining anomalies in other extinct craters. After the subsidence, the eruption must have been sustained for some time, for the ash above the lava is upwards of 100 feet thick, and dips away all round from the Blue Lake, showing *that* as its centre. I do not think that the subsidence was caused by the eruption of lava from the first and most ancient crater, — the lava would have then been heated enough to make it plastic, — but it appears to have been perfectly cool when broken, and the fractured edges are sharp and jagged. The subsidence more probably took place when a subsequent eruption had caused an underground flow of lava. This of course would have been larger in quantity, and would have given rise to a larger vacuum.

The eruption, then, was this:—The boiling lava, from whatever cause arising, may have pressed heavily against the overlying strata, so as to crack and fracture it in many places. The pressure which would force a mass of rock half a mile wide, and in thickness equal to the depth of the lake, at least 240 feet, must have been enormous, and this, when exercised on the soft friable rock of the sides, or, it may be, on what is mere sand (underneath the coral strata), when combined with heat, would easily force a passage towards the sea; and once an outlet was obtained, the absence of lava would cause a hollow, and finally a chasm, through which the eruption of ash would have full play. To a failure of support, consequent on a subterranean outpouring of lava, I attribute the chasms of both the Blue and Valley Lakes, with this difference, that while the eruption continued throughout the whole extent of the Blue Lake after the falling in of the rock, in the Valley Lake it was confined to the west end of the chasm, formerly the high wall or peak of Mount Gambier. The eruption, then, of the Blue Lake was simply limited to the ejection of large quantities of ashes and occasional fragments of rock, continued for some time after part of the boiling fluid had made a way under the soft limestone rock, and flowed down to the sea.

We have seen that there are four extinct craters at Mount Gambier, besides the remnant of a fifth.

Some of these may have been in activity together. There were, I think, three periods during which the craters were more active than at other times; though the rests, apparently, were only temporary, and far from leaving the mount in a perfect state of repose.

The following are the reasons upon which these suppositions are based:—At a short distance from the lake the ashes are found to lie in three distinct layers, all composed of coarse tufa underneath, and fine ash-dust on the uppermost side. Each layer was doubtless caused by a distinct violent eruption, which, on the commencement, would scatter large fragments about, and, as the energy subsided, a fine ash-dust would gradually cover them over. The eruption again breaking out, would renew the large fragmentary layer, thus marking its own periods of disturbance by distinct strata.

That the lulls were only very temporary may be seen from the fact, that the fine dust on the upper side of the lower or of the middle strata had not time to become the least altered before the second and third deposits were superimposed. Generally, above the upper layer there is a mass of rich black loam, covering it at a variable thickness. This is decomposed ash, originally of a fine and, therefore, easily decomposable texture. This latter deposit is easily accounted for, because, after an outbreak, there ensues in all volcanoes a long period of quasi-disturbance, during which time the eruption is, as it were, settling down, and the crater cooling. In this interval smoke and fine dust are continually emitted, and cover the ground to some extent in those places nearer the crater.

There remains now only one point to be noticed, and that is as to what geological date we are to assign the period of disturbance. One thing only is certain, that it happened since the Crag period, though at what precise epoch there is no evidence to show. The fossiliferous rocks, so often alluded to, are of the Mount Gambier formation, described in former chapters of this work. The ashes, as before stated, are resting above them, and evidently there has been little or no upheaval since the volcano broke forth. They were formerly, beyond a doubt, part of a coral reef, and immense masses of a common extinct pecten may still be seen in the walls, with lumps of coralline of the species alluded to,

Pecten coarctatus. Mount Gambier.

classified under the name of *Cellepora gambierenis*. Of the peculiarities of the strata, however, I shall say only one word more at present, so as not to repeat what has already been described in another part of this work.

Wherever the beds are found caves also appear, many of which (by piles of bones, &c.,) are seen to be certainly not later than the Post-Pleiocene. The rocks, therefore, were in that period in the position they are at present, which they were also in when the volcano broke out; so that if hereafter caves should be found with ash, &c., inside, or bearing marks of having been disturbed by the eruption, some better approximation may be made to the geological date: but at present the rocks cease to guide us farther.

But was the volcano in activity lately? An answer in the negative may safely be given, because, first, the ash is quite decomposed in many places, and the porous lava partly so, which must have taken considerable time to effect; and next, the large crater is filled with water to the depth of 240 feet, which water could not even have begun to collect until the rock was perfectly cool, and then must have taken ages to become the large body of fluid at present resting there. To give an idea how long it takes volcanoes to cool, or for the ash to decompose, I will mention a few instances. The lava of Jorullo, which poured forth in 1759, was found to retain a very high temperature half a century after. The ashes on the Peak of Teneriffe are nearly undecomposed, and yet it is not known to have received any fresh additions during the last 600 years. Some of the ashes on extinct volcanoes in Auvergne, which I visited in 1853, are much less decomposed than those of Mount Gambier, and yet the former have been deposited more than 1,800 years ago. Now, when it is remembered that the ashes of the latter are not only decomposed, but that large trees have taken root and grown up in it, we must be of opinion that our volcano has been extinct for some considerable time. And let it be remarked, that the cases I have mentioned are not exceptional, for I could specify many more, which would all bear testimony to the antiquity of the mount.

When, however, we say that it has not been in activity lately, there is no intention of asserting that it is impossible for it to break out again: tranquil as it may appear, the igneous agent may still be active below. It should not be forgotten that Vesuvius was quite as tranquil about eighteen centuries ago. Indeed, when reading the description of the former state of Vesuvius, its great basin, in which trees and grass grew, and an army was once encamped, one is forcibly reminded of the present state of Mount Gambier.

If Vesuvius has become what it now is, Mount Gambier may yet do so likewise. At any rate, it is not completely at rest, for shocks of earthquakes have been occasionally felt, while the land around is daily upheaved. This latter fact is significant. No active volcano has been found otherwise than in the vicinity of land in the course of upheaval, though the converse of the fact hardly holds good.

I forgot to mention that there is always found between the ash and limestone, when at any distance from the craters, a thick bed of fine sand, showing that, after the upheaval of the reef from the sea, it became a sandy desert previous to the igneous outburst. Whether this sand supported any vegetation, or whether there was any vegeta-

s

tion in the surrounding country prior to the breaking out of the volcano, it is difficult to determine: none has been found between the ash and the limestone. I may mention, however, that I have seen fragments of scoriæ enclosing pieces of charcoal.

The minerals found in the craters are few, chiefly confined to olivine, with darker crystals of the same mineral embedded. The aborigines use the dolerite as a weapon, fixing it in pieces of wood, and forming a kind of axe; and, singularly enough, the same mineral serves a similar purpose to the Indians near the Cordilleras of South America.

I think I have now gone through the principal features of this curious volcano, in which I have often been obliged to sacrifice, for succinctness, many details I could have wished to have mentioned. We are told there is a philosophy in stones, and it certainly is strange what a history of the past a few rocks can give us. There has been a coral reef, a desert, and a burning mountain where beautiful lakes now rest, and each period has erected monuments to its memory. There is a history, too, written in plain characters, for the mind of man, and my occupation has been to decipher it.

Going back, in imagination, to the time when the coral was alive and covered by the sea, who could have thought it would come to be what it is now? But imagination is not needed. We have only to glance at the remains before us to realise the truth of the tale they tell. These rocks were once covered by the green waters. There, while the rising tide dashed its sparkling waves through the groves of coral, where the busy polypi were plying their variegated arms in search of matter to add to these structures, a thousand fishes frisked for a while to die and leave their forms imprinted on the stone, while the cunning saurian slept among the arborescent forms, or wilily watched his prey. Then the earth slowly raised them from the waters, and life faded away. Fishes and reptiles are gone, and stones tell how they lived and died. The reef became a sandy desert, without a drop of water or a sign of vegetation to relieve the eye—a vast and dreary solitude. But Nature soon changes the scene. Subterranean thunders are heard; earthquakes rumble and rock the ground. Then masses of stone fall in and give vent to smoke and steam, which rush from the centre of the earth. By and by, fire begins to appear, and Nature, no longer able to restrain the ravages of heat, sends it forth into a bubbling hissing cauldron of molten stone. Standing upon the brink (if human being could stand alive on such a place), while the air is darkened with smoke and ashes, and huge fragments of stone are being hurled into the air to fall into the hissing seething mass below; while the light from the fire and the noise of explosion blinds the lightning or outbids the thunder overhead; while the bellowing and splashing of a lake of fire make a scene at once horrible and magnificent, one could almost imagine oneself on the bank of Tartarus.

But comparison would be vain; not even Vulcan could stand and describe such a scene. He might have thought,

> 'In Chaos antiquum confundimur. . .
> . . . Neque enim tolerare vaporem
> Ulterius potuit, nec dicere plura.'   Ov. *Phaëton.*

But now how changed is the scene! the smoke has cleared away, and the fires are extinct. Nature is at her repose. The melted walls have cooled, and an azure lake covers them. The ashes on the bank are covered with verdure, and reeds grow where fire glowed. The underground thunders are indeed heard no more, but the wind sends a soft moaning through the shrubs, while the gentle splashing of the calm and glassy lake is now the only echo that is heard from shore to shore.

CHAPTER IX.

VOLCANOES—CONTINUED.

MOUNT SHANCK.—DISSIMILARITY OF VOLCANOES.—IMPORTANCE OF DESCRIBING THEM.—DESCRIPTION OF THE COUNTRY.—WELL-SHAPED HOLES.—VALLEY.—AUSTRALIAN FLORA.—SMALL LAKE.—VOLCANIC BOMBS.—THE GREAT CONE.—REMAINS OF FORMER CRATER.—HOW MORE RECENT CONE WAS FORMED.—ITS APPEARANCE AND SIMILARITY TO VESUVIUS AND ETNA.—INDENTATION IN THE SIDE.—EVIDENCE OF FORMER PEAK.—LAVA STREAM.—CURIOUS MODE IN WHICH IT IS HEAPED.—DERIVED FROM OLDER CRATER.—CAUSE OF HEAPING OF THE SCORIÆ.—PARALLEL INSTANCES.—CONNECTION OF MOUNTS GAMBIER AND SHANCK.—CONCLUSION.

AFTER having given my readers a lengthy detail of the extinct volcanoes of Mount Gambier, we now turn to the volcanic monument next in importance in this part of South Australia, namely, Mount Shanck. It is scarcely so interesting as Mount Gambier, being neither so extensive nor so varied; but it is important, as showing how far the views on the subject of the former crater are realised in this. One would think there was a great sameness in the character of volcanoes, because, having all resulted from the same cause, namely, the outburst of molten fires from the interior of the earth, the same appearances might be expected. On the contrary, however, there is the greatest variety. No two ever resemble each other, except, perhaps, in the conical outline and the basin in the centre, and it is in the description of their various peculiarities that so many facts connected with their history have been brought to light.

Lest any should think that in the following pages too much space is given to detail, it should be remembered, that even if the facts are new they are important, and may help to settle points in a matter where very little certainty prevails. Apart, however, from confirming a theory, the history of any volcanic phenomena cannot fail to be interesting; if it only should give us an idea of the extent to which our continent has been disturbed by fiery agency, before becoming a resort for the European, it would be well worth consideration. But it does more. It is a part of the history of the earth,—one of the many testimonies which the rocks bear to the wondrous structure of the ground beneath our feet—to the greatness of that Omnipotence which can let fires flow forth so as to melt rocks and rend mountains, and then seal them up so that flowers shall grow peacefully where they rose.

Another reason may be added for multiplying the records of volcanic action. We are far, even at the present day, from understanding the cause of volcanoes. Theories have been propounded, but uncertainty prevails. From Nova Zembla to New Zealand they are constantly met with, and though at this moment they are burning amid the snows of Iceland, the waters of the Mediterranean, and the heats of the equator, their origin and the manner in which they burn out are equally mysterious. In this state of things, the accumulation of records is of great importance. Every little (in which category these observations are included) may be of use.

It will be remembered, that in explaining the geological features of Mount Gambier it has been stated:—1. That the lava arising from the eruption has, in all probability, flowed underground. 2. That the eruptions do not appear to have given rise to any upheaval or elevation in the immediate neighbourhood of the walls; on the contrary, subsidence seems to have been very frequent. 3. That to sudden subsidences of small areas are to be attributed some of the Mount Gambier craters. It is necessary that these particulars should be borne in mind, because they are elucidated by what is now to be described, and because, as Mount Shanck is between Mount Gambier and the sea, some confirmation must be looked for of the fact (if fact it be) that the lava flowed underground.

Mount Shanck, as seen from Mount Gambier, appears like a truncated cone, rising abruptly from an apparently level plain. There are no mountains tending to break the suddenness with which it appears on the field of view, and its darkened outlines readily suggest to an observer an extraordinary origin. The country around is well and almost thickly wooded, the general aspect being fertile and pleasing, even seen from a distance. It is about eight miles, or even less, from Mount Gambier, the sea being about ten miles farther on.

Enough has been said in the last chapter about the latter mount; but, in taking leave of it, I cannot refrain from mentioning the very beautiful view that is to be obtained from its summit. Below, the Blue Lake, with its smooth dark waters, and, a little to the north, the white houses of the township peeping out amid the trembling branches of the trees; all around green patches, which wave more and more in the breeze as the harvest approaches; whilst many a curling column of smoke, or the echoing of the whip in the forest around, tells that the new colonist is making a home where industry has never toiled before. This is the picture immediately around.

In the distance, to the north-west, Leake's Bluff rises, while the outlines of Mount M'Intyre show more dimly on the sight, and then a thin blue line, extending from the west to nearly south-east, shows where the ocean limits this part of South Australia. Mount Gambier is not very high, but the country is so uniformly level that a very small elevation gives an extensive field of view.

Descending from it, and making for the cone with which we are at present occupied, one is astonished at the rich, the meadow-like appearance of the country. After being out some time in these colonies, we become used to a certain dried-up appearance in every landscape, and learn to forget the flowery pastures which used to meet the eye at home in the month of June. At the sight of the country at this mount the old ideas come back with vividness. There is meadow land as thickly studded with the buttercup and blue-bell as the finest hay-field at home. 'Beautiful' is an adjective which comes short of the reality; and it may be doubted whether Somersetshire, or Kent, or Leicestershire could produce finer meadow land than the country between Mount Gambier and Mount Shanck. Alas! that this should be a rare exception in South Australia.

There is rather an extraordinary thing common in the rocks about half-way between the two mounts: these are well-shaped holes in the ground, close to each other, and though they descend perpendicularly, no bottom can be found. One is about a yard in diameter, others being less; and through the moss which covers the sides one can easily see that the fossiliferous limestone has been bored through. If any solid substance is dropped down, it can be heard rumbling for some distance, the noise growing gradually fainter till it dies away, but no stoppage of any kind can be detected. Supposing the lava to have passed underground in this direction, it would not, at first sight, seem unreasonable to attribute their origin to steam arising from the melted liquid: such is the opinion of the people here. But this, perhaps, is too easy a theory to be the correct one. They may be accounted for like the sand-pipes in the chalk; but as they occur in other places where there is little probability of the existence of lava, they could scarcely have arisen from steam. They have been already alluded to.

As the volcano is approached, the ground becomes broken and very hilly. The soil, too, is less rich, as evidenced by the quantity of stringy bark (*Eucalyptus fabrorum*) and grass-tree (*Xanthorrhœa australis*). The ferns (principally *Pteris esculenta*, *Asplenium laxum*, and *A. flabellifolium*) and underwood also become thick and intricate. By and by, large blocks of porous lava are seen strewn on the ground, and a peculiar brown ash-dust rises under one's feet on crossing the numerous abrupt spurs which run out from the base of the cone.

On ascending the steepest of these, a dense tangled mass of vegetation comes into view, evidently surrounding some hollow below. Descending towards it over treacherous and steep ground, hidden by brush, and taking a sweep round, to make the descent more easy, the bottom is reached, and a pretty little lake comes in view. Situated as it is in a kind of dell, it wears a most silent solitary aspect. The lava boulders and limestone rocks, however, jut out from the dense and high brush-wood in black and white patches, the occasional slopes of silver grass gracefully interrupt the thicket, and, with the help of the trees which hang their branches around, the loneliness is turned to beauty.

It is one of those places where the beauty of the

Australian flora can be seen to best advantage. The tall dark tea-tree (*Melaleuca paludosa*) reflected in the smooth water, the ferns and mosses making a carpet underneath the mimosa; the *Bursaria spinosa*, and *Calycothrix scabra* with its bushy pink flowers, filling up the interstices in the brush; and the whole united by the delicate tendrils of *Comesperma volubilis* with its network of blue blossoms; make a scene as beautiful in its kind as the vineyards of Provence or the rich palm scenes of the torrid zone. No better idea could be formed of this little place than from Sir W. Scott's description of that dell whence issued the skiff of the 'Lady of the Lake.' Had South Australia been long enough inhabited, this spot would, perhaps, have been invested with traditionary legends, making the mount the scene of wild incantations and the resort of fairies.

This little lake is just at the foot of the cone, in fact, almost situated in the side of it, and has arisen from a small eruption which has proceeded from its centre, probably at the time Mount Shanck was in activity. Very little lava or ash has come from it; of course some has come, and the sides being near the point of eruption, the ejectamenta deposited immediately after their egress in a partially fused state have formed layers of scoriaceous lava, which evidently commenced flowing back to the crater before cooling. This is all that is found. In some places the trap lies in layers just as it cooled, and in others it is broken up into boulders.

It is very porous, but more so on the top of the layer than underneath; the latter fact is easily explained. The pores owe their origin to the escape of gases from the melted fluid, and these gaseous bubbles would naturally rise to the surface, but as the portion exposed to the air would cool first, a cake would be formed on the outside which would prevent the exit of the bubbles, and so they would remain, after cooling, in the shape of almond-like holes in the stone. Those who are familiar with the interesting narrative of the voyage of the *Beagle*, will doubtless remember the description of the volcanic bombs found, I think, at Ascension Island. Their close vesicular structure is explained by supposing the outside to have cooled while the interior was still in a state of fusion. This is somewhat similar to what has been just said. The same phenomenon has been used by Mr. Henessy in illustration of his theory of the slow cooling of the earth.

At the north end of this basin and on the east side the limestone is not covered with ash, but stands out in small escarpments, even far above the level of the lava. It is blackened and was not formerly fossiliferous, being of the uppermost limestone strata, which in this formation seldom contain many shells. Probably, the reason why it stands so far above the water-level, and in broken masses, is because the spot was disturbed by a slight earthquake at the commencement of the eruption, and its perpendicularity explains why it was not covered with ash or lava. It is evident that this lake was not so much an eruptive crater as a spot whence issued steam and a small quantity of ashes.

These are not unusual in volcanoes. There is one by the side of Vesuvius, which, though it sends forth ashes occasionally, confines its operations nearly entirely to steam; there is another by the side of Etna. What is the cause of them, or why the main crater is not a sufficient outlet for the steam, is not well understood, though, when they occur independently of volcanoes (such as in the case of the suffioni, in Tuscany), an explanation has been readily found. The one now described is certainly a supplementary point of egress; for, had it been a proper crater, it is quite large enough to have given rise to a very large quantity of ash, whereas, at present, the walls do not rise above the plains.

At the side of this lake the cone of Mount Shanck rises abruptly. The ascent is very steep, and, though covered with thick grass, is only scantily supplied with trees. Occasionally, a broken fragment of porous lava is met with, but with these exceptions, which are rare, the sides are smoothly sloping. Going to the top of the cone (no easy matter, for the inclination is enormously steep, and the height 500 feet), you stand on the edge of the crater. It is a deep dark abyss, the walls around forming a complete circle of almost equal height. Its aspect is entirely different from Mount Gambier, though quite as grand. There is no water at the bottom to give it that air of placid loneliness which the other possesses; but the dark stone walls, occasionally covered by a verdure which the shade makes darker still, the suddenness of the descent and the yawning look of the chasm give it a wild sublimity, grand and awful of its kind. The whole depth of the crater does not probably bring it much, if at all, below the level of the limestone strata.

The shape of the basin is oblong, and the western side the highest. In this particular it resembles Mount Gambier, but there is not much difference between the highest part and the rest of the walls, the edge, though broken, being pretty equal. On the west, or highest side, the descent to the bottom is more precipitous and sloping than on the eastern. In the latter case the sides slope down half-way to a kind of platform, and then descend in broken undulations to the bottom.

On the exterior of the mount, at the west side, there are the remains of a former crater. It is just a half-circular wall, joining on to the present cone. Its form, however, will be better understood by explaining how it has been changed to the state in which it is now seen. Supposing the circle of ash to have been once complete, of which fact there can be but very little doubt, the point of ejection must have been in its centre. When this point cooled, and the eruption had ceased from that part, another broke out right in the middle of the eastern wall. This, after breaking away all the wall which was over it, deposited its ashes, scoriæ, &c., in a circle round its point of ejection, thus cutting the old crater in two, and, perhaps, taking the materials of the side where the second eruption broke out to form new walls.

The most perfect cone is, therefore, the most recent, and it has certainly been the most extensive. The old crater is not more than half its height. The sides of it are steep, both in the interior and externally. They are, apparently, more loose and more decomposed than the newer crater. The ash seems a white powdery tufa, with fragments of felspar, porphyry, and scoriæ embedded. There are large trees growing both on the outside of the wall and in the basin, but none on the side formed by the newer cone. Tall gum trees are common; and this fact is the more remarkable, as there are no *Eucalypti* to be seen on the inside of any of the Mount Gambier craters, and there are no trees at all on the inside of the Mount Shanck cone. Though shrubs abound, being newer, and the ash less decomposed, there is, probably, no soil in the latter of sufficient depth to support them. The side of Mount Shanck which slopes up from the old crater is so fearfully steep as to be almost precipitous, and it would be almost impossible to ascend the mount from that side.

Returning to the top of the walls of the higher cone, and looking into the basin, the sides are seen to be composed of regular layers of ash, which have hardened into a vast conglomerate, like the higher wall at Mount Gambier. In some cases, there has been a backward flow of the lava which has bubbled out. It appears twisted into strange wreaths, like the gnarled roots of some huge tree. The sides are nearly entirely covered with vegetation, except at the top, but there are places where the black ash is undecomposed, which do not bear a sign of vegetation from the top to the bottom. Looking down the crater is exactly like looking into a large funnel, so very narrow is the bottom in proportion to the rim. There is no break in the side, nor outlet of any kind for lava; in fact, the whole process of the eruption seems to have been limited to throwing up masses of ash until it had burned itself out.

It is interesting, however, to know that the state in which the crater is now seen is probably the appearance it wore (with the exception of the vegetation) at the time of its activity. When Sir Humphry Davy visited Vesuvius, he says that whenever the smoke cleared away, and he could look down into the crater, there was no fire to be seen, but it appeared like a deep black funnel coming to a sharp point, from which smoke and steam were rising. Every now and then, a noise was heard like distant thunder, which, coming nearer, seemed to end in an explosion at the bottom of the crater, casting up volumes of ash into the air, and then all was quiet again. The crater of Mount Etna, as described by Sir William Hamilton, seems to be just similar,—a dark funnel, with no fire visible, casting up ash in an occasional explosion.

Thus it appears that both resembled, in their quiet state, Mount Shanck's present aspect.

There is one peculiarity in the layers of ash which is worthy of notice. On the western side, close to the higher wall, there is a deep indentation or notch in the lip of the crater. This appears to have been made, after the deposition of the side, by some explosion or other violent cause, because the strata are seen to be sloping from the bottom up to this spot, that is, dipping away on each side from the indentation. This shows clearly that the part in question has been a high point in the side of the cone. Indeed, it would appear, from the whole appearance of the interior of the crater, that this place has been somewhat similar to the highest peak of Mount Gambier, for the general bearing of the strata of ash conglomerate is towards the point where there is now only one indentation at the summit. There must have been an elevation there originally. Probably there was a peak, but the walls being too narrow and steep to afford a good foundation, it toppled down into the old basin, which is just on the other side.

It has been said that there is no outlet for lava visible, that is, that there is no side of the crater wall broken down for a lava current. On the north side there is a very distinct stream of lava. It comes directly from the side of the walls in a high heap of scoriaceous fragments, and then, instead of continuing in a regular stream, is traced onward by a succession of hillocks, the first three of which

T

are upwards of twenty feet high, with very little elevation of lava between them. It makes a rapid curve to the southward, and after about fifty yards divides into two or three separate streams, still preserving the same uneven outline, only that the hillocks are much smaller, until the stream spreads itself over the surface and becomes altogether lost in the course of about half a mile.

During its course, the ground appears very much broken. When it is said that the ground is broken, it is meant not only broken up into hillocks, but also covered throughout its length and breadth with fragments of scoriæ, from one to three feet in diameter. Sometimes these boulders are gathered into mounds, as though piled up by art, and again they are found lining a deep hollow; but whether rising into hillocks, or scattered about as if thrown from one centre, they all keep a regular line, at times diverging from due north and south, but only to make a slight curve and then return to the original course. It appears, in fact, like the course of a liquid, and this was really the case.

It seems, for many reasons, very clear that this stream proceeded from the ancient crater, and not from the more modern one. In the first place, such a stream could hardly have forced its way from underneath the walls and not have caused them to give way above, or, at all events, to have shown that they had been subjected to some pressure. But, on the contrary, the outline is quite unbroken, and does not at all appear to have been pressed upon from underneath. If, however, we suppose the lava to have proceeded from the old crater, and that the ash was subsequently deposited on the top from the second eruption, we can easily understand the appearances. Again, the view of the interior does not at all convey the idea that the lava had proceeded from it. But the elevation on the eastern side at the bottom of the basin seems like a part of the old stream, the rest of which has been destroyed by the breaking out of fires underneath.

Many have imagined that this lava stream, from the fact of being piled up so irregularly in heaps, is nothing but the scoriæ which has been derived from the crater; but, on examining the ground, it will be seen at once that the scoriæ could not have arisen from the adjacent crater, because it takes its origin close to the north side of it, and then runs along in an undeviating line till at least half a mile past it. It need scarcely be further stated, that, if it came from the crater, it would be scattered at least half-way round in a semicircular form, the larger fragments being generally nearer the cone. But it is not so. There is a regular straight line nearly north and south, occupying only one side of the volcano, and pursuing its course quite independently of it.

This line of lava was, then, a current from Mount Shanck's ancient crater; but, in supposing it to be so, it is not easy to account for the broken undulatory character of the scoriæ. It has been stated that the pieces were piled up together, and sometimes seemed to surround hollows in the ground. This state of things could have been produced in two ways:—The first is, by supposing the upper crust of lava to have been heaved up, after cooling, by a new current running underneath. This would raise the stone almost upright in slabs, and probably, if they broke afterwards, would form the piles of scoriæ which are seen. But some of the piles are over twenty feet high near the point of eruption. This might arise from the comparative coolness of the lava, which would make it flow slowly in a very thick stream. To bear this out, the following passage from Wittich is cited:—

'There is probably no other liquid matter which is possessed of such a degree of cohesion as running lava. We must come to this conclusion when we find that this matter does not spread over the inclined plane down which it runs, but forms a ridge having exactly the shape of an embankment, or a rampart with regularly sloping sides. The ridge is commonly of considerable height. Even small streams of lava are found to rise from ten to twelve feet above the adjacent ground. Larger streams are sometimes from forty to fifty feet high. The lava which issued from Skaptaar Jokül was at some places from ninety to a hundred feet above the ground over which it had flowed.'*

A second cause of the piling of the stream might be the explosion of disengaged gases, where the upper part of the current had cooled and the under

* *Curiosities of Physical Geography.* London: 1855.

was still flowing. In illustration of this, a further quotation from the same author will be pardoned:—'Occasionally a very loud report, similar to the firing of a cannon, is heard to proceed from a stream of lava. This happens when the lava runs over a swampy ground or a very moist soil. The sudden conversion of the water into steam, and its decomposition, produce a commotion which for some moments is able to stop the progress of the stream. The stream breaks with great noise through the mass, tears asunder the crust of scoria which envelopes it, and throws both the lava and scoria into great confusion. As a portion of the stream is decomposed, the hydrogen explodes, and produces the loud report above mentioned and the accompanying flash.' This would be more likely to happen when the lava first touched the ground, and consequently near the crater, where most disturbance of the lava stream of Mount Shanck is found.

The whole thing may, however, have arisen from the manner in which the lava flowed. Most observers who have had an opportunity of witnessing volcanic eruptions, such as Sir W. Hamilton, Dolomieu, Dr. Clarke, &c., have stated that a flow of lava generally moves (when cooled to a certain extent) in large uneven sheets; but this refers to localities where the flow is very extensive. In those places where the ejected matter is small in quantity and only molten in the centre, the stream (according to Mr. Scrope, quoted by Lyell) is like a huge heap of cinders, rolling over and over as it

went onward. The following is the passage :—
'The surface of the lava which deluged the Val del Bove (Etna) consists of rocky angular blocks, tossed together in the utmost disorder. Nothing can be more rugged or more unlike the smooth uneven superficies which those who are unacquainted with volcanic countries may have pictured to themselves, in a mass of matter which has consolidated from a liquid state. Mr. Scrope observed this current, in the year 1819, slowly advancing down a considerable slope at the rate of about a yard an hour, *nine* months after its emission. The lower stratum being arrested by the resistance of the ground, the upper or central part gradually protruded itself, and, being unsupported, fell down. This, in its turn, was covered by a mass of more liquid lava, which swelled over it from above: the current had all the appearance of a huge heap of rough and large cinders, rolling over and over, chiefly by the effect of propulsion from behind. The contraction of the crust as it solidified, and the friction of the scoriform cakes against one another, produced a crackling sound. Within the crevices a dull red heat might be seen by night, and vapour arising in considerable quantity was visible by day.' *

Now, it will be observed that, in the case we have to consider, the flow of lava was very small, and therefore must have solidified very shortly after its

---

* Lyell, *Principles of Geology*, 9th edit.; see also Scrope, on *Volcanoes*.

emission from the crater, and so probably, as the heaps are larger near the crater, and smaller as the stream is followed on, it is because as the first became cool, and was rolling over in heaps, fresh lava flowed underneath, and so raised them higher and higher.

And now, as to the question whether Mount Shanck is in any way connected with Mount Gambier. Let us first suppose that the lava of the former flowed towards the sea underground, a supposition for which reasons have been given in a former chapter, would the mere underground flow of an immense fluid mass of fire give rise to a volcano like Mount Shanck? Very likely the obstruction of an underground flow of lava, which would cause a large igneous subterranean lake to collect, would, by its bubbling and seething, give rise to a sort of crater, just as fissures in a cone form a lateral crater. We may safely, however, answer in the negative in this case. Certainly Mount Shanck does not appear so large as to have been a lateral crater, but then its distance from the other mount, and the fact that there is evidence of several separate eruptions, point out two distinct foci of disturbance.

The connection between these two extinct craters was of a deeper origin. They both belong to some great area of disturbance, which not only connected them, but also the volcanoes. to be described in the next chapter. Probably no two of the craters of either of the mounts were in activity at the same time, because, as they must be regarded as vents or safety-valves, by which the pent-up fires underneath sought a relief for their steam and gases, it is difficult to imagine that one point of eruption would not relieve localities so near as these two cones. This is, however, a matter more of connection than fact.

It has not been mentioned, that for four miles round Mount Gambier the country is very hilly. These elevations may have been caused by earthquakes which preceded the eruption. An examination of these hills might be very interesting, as showing the way the earth-waves were transmitted, and what was the extent of the shock and manner of the disturbance. Very little disturbance, apparently, took place round Mount Shanck, though the country is slightly hilly in its immediate vicinity, and one or two circular pits occur.

It may also be stated, that lava from the second eruption of Mount Shanck may have flowed to the sea, because the trap which is found on the coast is directly to the south of the latter, and, consequently, to the south of Mount Gambier, which is almost due north. No difference in the composition of the trap which is found on the coast can be traced, even so much as a separation of the strata, for it dips rapidly into the sea, and very little of it can be seen. Whatever other traces exist on the coast is difficult to say; for the sand, as before remarked, is drifting up so fast that even trees are buried in its encroachments.

The coast line, as seen from the mount, is barren

and dismal enough; but on a closer view it bears so wild and lonely an aspect as almost to make one shudder. Large and dreary swamps covered over with dank vegetation, white sand-hills bearing patches of salt bush, and cold and gloomy cliffs are all that meet the eye; while the sea breaks in with such a heavy surf, that even in calm weather its solemn roar may be heard for miles around. It is seldom visited by human beings; and when a vessel was wrecked there some time ago, the dead bodies of the poor creatures who escaped drowning had fallen to pieces in the rigging before their remains were discovered. Forlorn and sad as it is, nothing could be more in keeping with such solitude to think that here volcanic fires rolled in times gone by. A long time ago it must have been — how long, indeed, may perhaps perplex mankind till time shall be no longer. A thin seam of shells in the sands, far above the water, tells us that even the sea has retreated since then — that the waters now surging at a distance were once beating their monotonous music on the spot where we can now stand, bringing into competition the noise of fires and the rush of waters. The stones, so full of strange histories, tell us that it is a long time since the fires rose; and the trees and flowers, quietly growing on the softened rocks in the crater itself, tell us, by their tranquil growth, that the fire has long since fulfilled its Author's work and disappeared, leaving for ages the black and empty chasm staring into the heavens, lonely and desolate.

CHAPTER X.

THE SMALLER VOLCANOES.

SOUTHERN END OF THE DISTRICT ONLY VOLCANIC.—LAKE LEAKE.—LAKE EDWARD.—CRATERS OF SUBSIDENCE.—LEAKE'S BLUFF.—MOUNT MUIRHEAD.—MOUNT BURR.—MOUNT M'INTYRE AND MOUNT EDWARD.—LINE OF DISTURBANCE CONNECTED PROBABLY WITH VICTORIAN CRATERS.—PERIOD OF THEIR DURATION, AND THE TIME WHICH HAS ELAPSED SINCE THEIR EXTINCTION.—SUBMARINE CRATERS.—JULIA PERCY ISLAND.—CONTROVERSY ON CRATERS OF ELEVATION AND SUBSIDENCE.—BOTH APPLICABLE HERE.—TRAP NOT ALWAYS CONNECTED WITH GOLD.

IN the last chapters, we have been occupied in examining those extinct volcanoes of this district which rise in the form of cones to such a height as to entitle them to be described as mountains. In this chapter we have still to do with craters, but in the form of lakes, and with volcanic phenomena which are neither cones nor craters, but dykes or faults.

It has been already mentioned, that the volcanic disturbance of this district has been entirely confined to the southern end, and the distance between any of the craters, even of the disturbed district, is so small, that if a line were drawn encircling the whole of it, it would not enclose very many square miles. The most northerly of the craters is Lake Leake. This is a large lake, about a quarter of a mile in diameter, very deep in the middle, but shallow on the edges, with reeds and bushes growing all round. The banks are very even, seldom rising more than ten or twelve feet above the water, except on the eastern side, where there is a sudden rounded eminence, about sixty or seventy feet in height. This is entirely composed of volcanic ashes, enclosing, here and there, small fragments of scoriæ. This hill slopes away pretty gradually on the side opposed to the water, but on the other side it is precipitous, descending very abruptly to the water's edge. With the exception of some black mud enclosing scoriæ in the banks all round, these are all the volcanic evidences of this place.

Close to it, however, and a little to the south, there is another crater (Lake Edward) rather smaller in dimensions, and having no eminence on its banks. This is also an extinct crater, as may be easily ascertained by a close inspection of the black mud which surrounds the edge of it. It is very deep in the middle. Both these lakes are twenty-two miles from Mount Gambier. They are rather singular volcanoes, and I am not aware that any parallel to them is to be found elsewhere. The fact of their being so wide and deep, showing that some very extensive igneous disturbance must have caused them, and yet to have given rise to a very little ash and no lava, seems exceedingly strange; in fact, they bear out the view already taken of the eruption of Mounts Gambier and Shanck, which makes the craters rather exceptional instances, or departures from the usual manner in which volcanic eruptions take place. Instead of these lakes being craters of elevation, they have been craters of subsidence.

On the banks of both lakes there are masses of limestone cropping out, occasionally showing that the strata were not upheaved, but the lake formed by a part of it falling in from volcanic disturbance underneath, and giving rise to a chasm through which ashes were cast forth. It appears very strange that an eruption which would cause such large openings in the surface should have been followed by so very small an amount of ejectamenta. Probably these chasms may have been caused by the void arising underneath the surface from lava that was pouring out elsewhere. There are two small hills, perhaps not more than 200 feet high, very near the lakes, and there is no trap rock visible upon them, yet their rounded outline and isolated position suggest a connection with the igneous disturbance below; moreover, their strong resemblance to hills to be mentioned subsequently, which are certainly volcanic, places their origin almost beyond doubt.

Returning, again, in a southerly and somewhat westerly direction, we come upon Leake's Bluff. This is a high bluff, as its name imports, raised, perhaps, 500 feet above the level of the surrounding plain, almost precipitous on its southeastern side, and sloping away quite gradually on the NW. The precipitous side is trap rock, very slightly vesicular, brown, compact, and ringing under the hammer. On the summit there is some of the *unaltered* coralline limestone, which has been tilted up by the trap rock, and it continues down the sloping side into the plain. There are a few irregularities in the slope such as would be caused by the upraising of such a mass of limestone and its doubling over on itself at the foot; beyond this, however, the country around does not seem to have been much disturbed, except in the direction of Mount Gambier, where hillocks occur like undulations of the surface, which become higher and larger until within two miles of the latter crater, where the country all round is disturbed, as if the eruption had been preceded or accompanied by earthquakes.

This bluff deserves some consideration. In the first place, it has given rise to a fault in the limestone strata; that is, the continuity is broken by the escarpments of trap rock, and the sequence of the strata which have been followed thus far must again be sought on the top of the bluff. On the other side of the escarpment the hill slopes gradually down, making the outline like the segment of a circle. I suppose that at the point of junction between the trap and the limestone, the latter would prove to be very considerably altered, but there is no opportunity afforded for an examination.

Both rocks are very much decomposed, being not only disentegrated and covered with grass, but also even with trees; but the limestone on the surface, which crops out at a little distance from the summit, is decidedly unaltered. The trap could not have been upheaved to its present position in a state of fusion, not only because the outlines of the escarpment appear clearly fractured, as though broken when in a state of solidity, but also because the general character of the disturbance is against such a supposition. It must have been then that a large quantity of liquid lava was injected under the limestone, and there cooled under the pressure of the immense mass of stone above and the force from below which injected it; a second pressure from below upheaved it bodily into the position in which it is now seen. It is impossible now to trace the extent of the fissure caused by this upheaval, in consequence of the manner in which the strata are decomposed and altered on the surface. Doubtless the fault continues some considerable distance on each side. Had this been a place where there were two distinct fossiliferous formations, more remarkable results would now be seen.

The strata containing fossils belonging to one period might be traced continuously with one older than itself, which had been raised to its level, but there is only one kind of fossiliferous rock here, and therefore no mistakes can arise.

North-west of Leake's Bluff is Mount Muirhead, which is a conical hill, with trap rock on the summit, and limestone on the sides all round. This is a trap dyke. It would appear that not only was the trap injected through the strata, but it flowed a little on the summit, for it is like a cap on the top, and very vesicular. The soft nature of the limestone strata caused it to yield a great deal to the pressure underneath before allowing the igneous matter to break through, giving rise to a dome-like appearance to the base of the hill. In other places, where the strata are very hard and compact, and lying near other eminences, which enable it better to resist pressure, the trap dykes make a clean cut through, without in the least raising the rocks in the vicinity, only pushing them a little back.

Mount Burr, rather more to the eastward, is a bluff like the one first described, with this difference, that it is perhaps a little higher, and has no trap rock visible on the escarpment. Perhaps it slopes away more rapidly on the north-western side, and there can be no doubt that it owes its origin to an upheaval somewhat similar to that exercised in the case of Leake's Bluff. The limestone crops out in immense quantities on the summit, and appears much broken, as though by pressure. Probably, the reason why the trap does not appear is because the limestone is either thicker and more abundant at this particular point, or the trap was injected from a more deep-seated locality.

Again to the north-west, is Mount Graham, a trap dyke, like Mount Muirhead, except that the hill is more rounded, and extends more like a ridge, in a north-west direction. On the summit the rock is very vesicular.

About four miles to the west of Mount Burr there is another mount, called Mount M'Intyre, and this continues on in a chain with a hill called Mount Edward, until close to Lake Leake. This is also a hill with limestone at the base, and trap rocks at the top. It is more prolonged than the other hills, and less conical, but in other respects it is just like them.

After Mount M'Intyre there are no more volcanic evidences. The range disappears into a limestone ridge, very little elevated above the plains. The distance between Leake's Bluff and Mount Graham is scarcely eighteen miles, the two others (Mounts Muirhead and Burr) lying between, at the distance of a few miles apart. There is little or no connection between them, not even a range of the most insignificant proportions. They stand, on the contrary, almost isolated from each other, only connected by the identity of their bearing from Mount Gambier.

No one can doubt that these numerous evidences of former volcanic disturbance in this district have been connected together. Not only is this to be inferred from their continuing in the same line, but also because they are always within a short distance from each other, and confined to the southern part of the district—the only part, indeed, where there is any volcanic evidence at all.

It remains to be asked, what connection can be

supposed to have existed between them? It is a well-known fact in geology, that volcanic disturbance is very seldom confined to one particular spot, either in the past history of the earth or in what is taking place on the earth's surface at present. One active volcano is very seldom found alone, and extinct craters are generally grouped together in what are termed volcanic districts. The only apparent exception to the rule is in marine craters. These are isolated at times, and then are either extinct, such as Trinidad, Tristan d'Acunha, or active, as St. Paul's or Graham Island. In the latter cases, the evidence that there has been no other disturbance is only negative, the depth of the sea around preventing any certainty as to the absence of other craters, from the difficulty of an examination. There are, however, many instances of extinct and active marine craters being grouped together in large numbers: the Galapagos Archipelago, which, according to Mr. Charles Darwin, must have contained upwards of 3,000 craters, all extinct, and the Azores, where a great many signs of activity still exist.

From the fact that these phenomena have always been found associated together, it has been inferred by many geologists that there are, or have been, in volcanic districts, under the upper crust of the earth, lakes or reservoirs of igneous matter, whose gases and pent-up forces sought egress in many points, perhaps at some distance apart from each other. These districts may be of immense size, such as the volcanic parts of the Andes, in South America, which occupy so immense an area, or of the moderate dimensions of the locality which I have just described. Now, I think there can be but little doubt that Mount Gambier, Mount Shanck, and the other places mentioned above, have belonged to one and the same area of volcanic matter, underneath the upper crust; and the general north-westerly bearing of the disturbance is due to the greater diameter being in that direction, or from a weakness in the strata tending to make a fissure more easy in that line than in any other. Probably this district was not an independent mass of fused matter, but rather an offshoot from one more extensive. At about fifty miles east of Mount Gambier, on the Victoria side of the boundary, there commences an immense volcanic district, which may be traced, with very little interruption, to Geelong (250 miles distant), by immense masses of trap rock and extinct craters of large dimensions. This kind of country extends considerably to the north of this line; and it is underneath the trap rocks thus found, at the junction of the Silurian slates and ancient granites, that the extensive Australian gold-fields are worked. This large tract of country has evidently belonged to one immense subterranean igneous lake, and the various craters which appear are evidences of the manner in which it has sought relief from time to time. It appears rather more ancient than the Mount Gambier district, though both have arisen in a very recent tertiary period.

If this is to be accounted for, I will give what appears to me to be a reason, though it is quite theoretical, and may be far more fanciful than real. After a long duration of activity, I imagine there was a period of repose, during which not only the trap rocks and the lava streams on the surface have had time to cool, but also the upper crust of the fiery subterranean lake itself. Supposing now a second, but less violent, period to supervene; the cooled crust and the overlying rocks might prevent an outbreak directly above, and, therefore, the fiery matter would find an egress at the sides, where the superincumbent strata was weak, or at the sea, where there was much less resistance to be overcome. This may appear a rather extravagant hypothesis; but in all probability the depth of the subterranean disturbance might have been very great, and the land on the Victoria side being much higher than about Mount Gambier, the actual resistance to pressure, whether by a cooled crust of lava or surface trap rock, may have been greater, on the whole, than the force required to send the lava in the direction of the latter place. This would account for our volcanic district being more recent, and the existence of submarine craters near it, which, though already described, will be here briefly noticed.

At Portland, as already mentioned, there is most distinct evidence of subterranean volcanic action. The strata of basalt underneath the Upper Crag, and on the south side of the bay, over the Lower

Crag, have been alluded to. These have, probably, proceeded from an extinct crater, which lies about one mile from the shore. It is called the Lawrence Rock,* and consists of a small flat rock, surrounded by a scattered reef forming a rough circle, and between it and the shore another long low reef, composed of scoriaceous masses of lava. The rock itself is stratified. The uppermost layer consists of decomposed trachyte, of a cream-white colour, containing disseminated crystals of mica. Underneath this there is a thick stratum of amygdaloidal lava, the base of which is black. The lime which is in all the vesicles is of a transparent waxy appearance. Under this, again, the stratum is a mass of tufaceous deposits, of a brown colour, loose and friable, and containing small fragments of scoriæ. There are no trap dykes near, unless some of the projecting rocks formed of compact basalt may be considered such.

It would perhaps be taking too much for granted to assert that the rock is the precise site of the volcano, though the semicircular reef near it seems to favour such a notion. At any rate, the nature of the strata, composed, as they are, of ejectamenta of such thickness, makes it probable that the source of them could not have been very far distant. There is an island, called the Julia Percy Island, close to Portland, but at such a distance from the land as to render it invisible, except on very clear

* Running through this rock there is a thick trap dyke. This may probably be the 'cooled chimney,' now forming a hard rock, which has resisted the sea better than the ash deposits around.

days. This is volcanic. I have not had an opportunity of examining it, but I am informed that it principally consists of compact basalt. It may have been another site of volcanic emanations, but this is only conjecture. It is evident, however, that the land has been upheaved considerably since the outpouring of these igneous rocks. That of the Lawrence Rock is overlaid by the crag, which could only have accumulated in a pretty deep sea; that of the island is overlaid in a similar manner. The evidence of this is discussed in another chapter, and need not be recapitulated here.

Whether or not this crater was subsequent to the eruption of Mount Gambier, can only be decided by a very minute examination of the locality. There is one evidence, however, offered here which it may be well to allude to. The disturbance of this district, as shown in the large number of extinct craters, must have extended over a long period of time. It might be possible that one or more of the craters which are at a distance from each other were in activity at the same period; but this can hardly be supposed of the craters which are close to one another, such as the different lakes at Mounts Gambier and Shanck. This is more especially seen by the manner in which the ash and lava are deposited, as already described. But not only are they, then, separate monuments of periods of disturbance, but they show that long intervals of rest intervened between them. Mr. Poulet Scrope states, that when a volcano has been so long at rest that the melted rock has had time to cool, the next eruption is obliged to make a new crater, because the solidified rock in the old chimney makes an irresistible barrier. The occurrence, then, of so many craters shows not only that the eruptions were distinct, but also were separated from each other at such an interval as at least to allow the old lava in the chimney time to cool and become solid. Remembering, now, the remarks which have been made on the slow cooling of volcanic products, and bearing in mind the number of contiguous craters, we can easily understand how long a period of disturbance this district must have witnessed. The length of the period can only be inferred by analogy. Vesuvius has been known at one time to be at rest for many hundred years, and then its eruptions have been at irregular periods, sometimes many years separated. Probably at Mount Gambier the different craters had time not only to cool, but to allow plants and shrubs to mature inside them, for charcoal is often found between the different layers of ash and basalt.

In describing these craters, and their mode of eruption, Von Buch's theory has very often been cited as to the crater-elevation theory. This may need some explanation. Some time ago, the geological world was much puzzled to decide between two rival hypotheses. One was Von Buch's theory, which supposed all volcanoes to have been what was termed craters of elevation. The theory, apparently, was formed rather to meet a difficulty than from any stamp of probability it wore in other respects. It was supposed that when lava and ash conglomerate were found on the side of volcanoes, in very highly-inclined beds, it was impossible for them to have remained there, had their inclination been so great when they were deposited. To meet this difficulty, it was imagined that all eruptive craters commenced their operations by forming cracks or fissures through the level surface, and outpouring ash and lava upon it; that, after this had become consolidated to some considerable thickness, the whole was uplifted by a subsequent convulsion and formed the cone. By this means, of course the beds became highly inclined, the inclination in proportion to the height. This theory met with universal approval, not only because it seemed to meet all difficulties, but because it was propagated by one of the most eminent European geologists, whose services to science cannot be too highly extolled. There are instances which seemed to favour the theory. The cone of the volcano of Jorullo was uplifted 1,000 feet in a single night. Other instances were also supposed to be furnished by the records of what had taken place in the eruptions of the Bay of Baiæ. There were, however, geologists who objected to the general application of the hypothesis. However well it might account for what had taken place in one or two localities, they said it was with difficulty reconcilable with what was observed elsewhere, and many facts were directly against it. Sir Charles Lyell was one of the first to object to its being applied as a part of the history of every volcano. He pointed out that at Mount Etna the trap dykes which had been injected in the earlier eruptions remain quite vertical even now, and that the Val del Bove showed more signs of subsidence than upheaval. He went farther, and tried to show, that even those craters (such as Palma) which Von Buch had personally explored, and declared to be craters of elevation, were not so certainly the result of such a process.

Little by little, like all theories which have been formed more to meet difficulties than suggested by facts, it has, of late, fallen much into disrepute: whatever truth there was in it, it was certainly too generally applied.

The object of mentioning this controversy is to point out how here it may be seen that there is probably truth on both sides of the question, though subsidence is far more common in volcanic phenomena than elevation. To the latter cause we must certainly attribute such hills as Mounts M'Intyre, Muirhead, Leake's Bluff, &c., as they appear to have been upheaved bodily from the pressure of trap underneath, during volcanic disturbance. But all the craters have had some subsidence in or near them, with, perhaps, the exception of Mount Shanck. It is needless here to repeat what has been said of Mount Gambier; Lakes Leake and Edward are both cases in point. But the principal difficulty about the inclination at which lava and conglomerate can lie must certainly be discarded, because many of the circumstances mentioned in the previous chapters show that such may be deposited and rest on very steep inclinations.

In taking leave of the volcanic features of this district, it would probably be well to notice the error of those who imagine the occurrence of trap rocks to be an indication of gold. Because gold is found underneath basalt (the blue stone of diggers), it is supposed that some connection exists between the two deposits. Now, the history of such formations is this:—Gold veins occur in rocks of the Lower Silurian age, which cropped out on the former soil of Victoria. These were decomposed by the action of water in creeks, or by weathering. The gold thus liberated became rounded by attrition into 'nuggets,' and deposited in the alluvial soil formed of decomposed rock. After these operations, and in no way connected with them, the land was overflowed by lava, and many creeks which were full of nuggets were thus covered over. Miners are sometimes much astonished at finding trees and fragments of pebble, rounded, underneath the blue stone they have penetrated. The former existence of creeks explains the difficulty. One of the richest gold-fields, perhaps, in the world is worked in the bed of an ancient creek thus covered over. This is the Clunes Mine, at Creswick's Creek, not far from Ballarat. To look for gold, then, because trap rock occurred, would be like searching for it in tertiary limestone.

Gold has never yet been found in paying quantities in South Australia, although there are doubtless numerous quartz reefs and other indications. But such signs prove nothing. We might just as well be disappointed because copper is not found in the metamorphic rocks of Victoria as well as those of South Australia. As yet, we know but a few of the reasons why certain minerals are always associated with certain rocks, and we must not hastily conclude, because we have the latter, that the former must infallibly follow.

It has been stated that there are some granite rocks found in this district. They occur in the bed of the Murray, and run in an east and west line across the desert east of that river. They occur, also, in small localities south of that line. They are huge rounded rocks, of red granite, of a very coarse crystalline structure. They have been mistaken for drift boulders, but they are, in fact, intrusive, and, though the Lower Crag has been perhaps deposited around them, they certainly belong to the tertiary period. Their line of elevation runs at right angles to that of other Australian mountains.

CHAPTER XI.

CAVES.

DENUDATION AND ITS EFFECTS.—CAVES IN GENERAL.—BONES IN CAVES.—CAVES MADE BY FISSURES.—HOW BONES CAME INTO THEM.—PARALLEL INSTANCE IN SOUTH AUSTRALIA.—COURSE OF RIVERS IN CAVES.—CAVES IN THE MOREA.—THE KATAVOTHRA.—THE SWEDE'S FLAT.—OSSEOUS DEPOSITS.—HOW BONES BECOME PRESERVED IN RIVERS.—CAVES WHICH HAVE BEEN DENS OF ANIMALS.—KIRKDALE CAVE.—BEACH CAVES.—PAVILAND CAVE.—AUSTRALIAN CAVES WITH REMAINS OF ABORIGINES.—EGRESS CAVES.—THE GUACHARO CAVES.—OTHER CAVES.—CONCLUSION.

AS most of the rocks described in the preceding chapters have been of a loose friable structure, and composed of limestone, it must naturally be expected that great portions of the beds have been removed, and that consequently, evidences of denudation will be found. Denudation may make itself manifest in many ways: either by removal of large masses of rock, so as to make breaks in the strata otherwise unaccountable, or by the rounding of outlines, or by leaving sharp pinnacles of the rock that has been spared (of which kind there are so many instances at Guichen Bay), or by the chasm caused by the flowing of rivers, or, finally, by caves, which owe their origin to various causes.

The latter kind of denudation is that with which we have to deal in this chapter. Caves are so common in this district, and so varied in their characteristics, that some detail will be necessary to describe them all. I mean, however, to devote this chapter to the subject of caves in general, and the various theories which have been proposed for their origin. Properly speaking, this should belong to a work on geology rather than the description of a particular district; but the interest of the subject will apologise for the digression, more especially as it will convey instruction directly elucidating what is to follow.

Caves are found in nearly every description of rock, but more particularly in two, and these from entirely different causes. These are trap rock and limestone, the former being generally the result of violent igneous action, and the latter infiltration of some kind. With the former we have not much to do at present; but, as instances of the kind of cave meant, Staffa may be mentioned. This is too well known to need description; but the regular crystalline form of the sides, and the nature of the rock of which it is composed, show that the mere wearing of water had nothing to do with its origin.

The other kind of cave is that which occurs in limestone, generally stratified, but in any case only where the nature of the rock is such as to admit of its being easily worn away by the action of water. They are of the most varied kinds and shapes, but admit of being divided generally into four kinds, as follows:—1. Caves which have arisen from fissures in the rock, and are therefore wedge-shaped crevices, widest at the opening. 2. Caves which face the sea-shore, and are merely holes that have been worn by the dashing of the sea on the face of the cliff. 3. Caves which open to the face of a cliff to give egress to water. 4. Caves whose entrances are holes in the ground, opening very wide underneath, and having the appearance of water having entered from above.

For convenience, these will bear the names of— 1. Crevice caves. 2. Sea-beach caves, or dens of animals. 3. Egress caves, or passages to give egress to subterranean streams. 4. Ingress caves, or passages caused by water flowing into the holes of rocks, and disappearing under ground. Caves of these four descriptions are found in nearly every country where the limestone rock is of any thickness. It makes no difference to what age the rocks belong, as these subterranean excavations are quite as numerous in the older strata, such as the carboniferous limestone, as they are in the modern tertiary.

As long as observations were only confined to occasional instances of these phenomena, each cavern, as it was explored, seemed to give rise to new features, and each was thought to possess individual peculiarities. Now, however, that observations have become more numerous, and opportunities have been afforded for comparing and collating the facts, several general points of resemblance have been observed between all. These are given at some length by Mr. Phillips, in his 'Manual of Geology,' and I shall give them here, adding such particulars as I have been able to collect elsewhere, from other works on the subject, or from observation. First, it has been found that nearly every cave possesses in some parts of its flooring, either embedded in stalagmite or in the dust accumulated therein, organic remains, either bones, shells, or even fragments of human art. In most cases, these remains were found to have belonged to extinct species of animals; and, when this fact began to be well known, and was found to hold good, almost universally, it was supposed that these remains bore a strong confirmatory testimony to the universality of the Deluge. But, in time, this view of the matter was abandoned. Apart from the fact, that bones resulting from the Deluge ought to belong to existing species, because the earth was repeopled with the animals destroyed thereby, two of every species destroyed having been preserved in the ark, it was found that in very few instances were bones found under the same condition.

Again, it was imagined that these places were all resorts for beasts of prey, who naturally look for such places of retirement, and would bring thither their prey. But this theory was, again, found not to have a universal application, because either the bones were all of animals too small to have chosen a cavern as a place of resort, or there were circumstances connected with the manner in which the remains were embedded which precluded such a theory. At length, it was decided, that though the fact was universal, the manner in which the bones became accumulated was different in nearly every case.

Some of these circumstances will be explained as we proceed; but it is worthy of remark here, that the osseous caverns are, perhaps, the only instances in science where totally different causes have combined to produce universally similar phenomena.

Another peculiarity noticed in caves has been that, 'whatever be the character of their floor, they assume, at intervals, along their length the appearance of a great fissure in the rocks.' Again, 'very few of these cavities in the rocks are entirely free, on their sides and roofs, from remarkable depressions and cavities like those produced on limestone by currents of water, or the slow-consuming agency of the atmosphere.' Many of them which now convey water are not encrusted with stalagmite, as the Peak Cavern, in Derbyshire.

This cave shows the effects of erosion by water so strongly, as to impress most beholders with a conviction that the whole was excavated by the running stream. We will now proceed to mention the different caves where the four varieties enumerated above are well exemplified.

With regard to fissure caves, the deposits in these are more easily understood. If we suppose large rents to be made in limestone, either by upheaval, earthquakes, or other causes, and these subsequently becoming connected with caves by the drainage of surface water, there is no difficulty in perceiving how bones may become embedded.

In the first place, animals might be easily entrapped, either by falling in by night, or during a sudden flight; or the water might bring down their bones from the surface drained by it during its course. A good instance of this kind of cave was discovered in a hole near Plymouth, which was being removed for stone for the erection of the breakwater. A large number of solid masses of clay were laid open, entirely filling the cavities in the limestone; these were connected with fissures in the surface, which were also filled with the same sort of clay. In this clay were found the bones of many extinct animals, including those of extinct deer, tigers, oxen, foxes, horses, wolves, &c. Where the surface of the cliff was exposed, the caves appeared to be, in nearly every case, connected with fissures reaching to the surface, and where this was not evident a connection might reasonably be inferred, in consequence of the identity of the deposits.

It may appear unlikely that animals would be entrapped into fissures in the manner I have described; but I can mention an instance within my own knowledge, which will quite bear out the theory. At the limestone ridge, about twenty miles east of Mount Gambier, and in the colony of Victoria, there is a small hill of limestone, rather more elevated than the rest. This is completely undermined with caves, which run in all directions. They never go very deep, and, consequently, have never much thickness of rock for their roofs: this causes many circular holes in the roof of the cave, which are perfect pitfalls, being covered round with long grass, which partially hides them, and having, in most cases, a clear descent of about twenty feet to the bottom of the cavern. In following the windings of one of the subterranean vaults, I came once, after threading through a very narrow passage, upon a chamber rather more spacious than was usual here. This was lighted by a round aperture in the centre of the ceiling. Immediately under this there was a heap of kangaroo bones, bleached, dried, and heaped rather indiscriminately together. All round the chamber there were bones of the same kind, scattered occasionally, mingled with sheep bones (a flock of sheep was kept in the neighbourhood), and the flooring, though occasionally covered with a loose dust, some few inches deep, was rapidly becoming embedded in stalagmite.

There could be no doubt that these animals were all precipitated from above, when either feeding or jumping too near the surface, and in a very short time this vault will have the appearance of a bone cave. This will afford a good instance of how animals may become entrapped by fissures. In long caves, which seem to have been the course of a stream, the cause of bones becoming embedded in the floor is not so easily accounted for. Generally, when the caves are those which have formerly been entered into by rivers, or caves of ingress, such as mentioned above, the water has ceased running into them, or they could not be explored, and therefore the fact of their being former passages for a stream is more or less a supposition. It is known, however, that rivers do continually disappear in countries containing much limestone, and that they sometimes flow underground for a considerable distance before again coming to the surface. Supposing, then, that they caused all the caves that are attributed to them, would they necessarily carry down bones and fill the passages with them?

In answer to this, I will give the observations of the gentleman connected with the French expedition to Greece, given in the 'Annales des Mines,' in 1833, and extensively quoted by Sir Charles Lyell, in his 'Principles of Geology.' It appears that in the Morea there is a great deal of limestone, known by its included fossils to be of the cretaceous period. There are regular rainy seasons in that part of Europe, which last during nearly four months, and, at this time, the land is perfectly deluged. Instead of running off by streams into the sea, the water falls, in most instances, into valleys, which are quite surrounded by hills. It does not collect in these, however. The valleys are surrounded with large fissures in the limestone, called, by the Greeks, *katavothra*, down which the water washes and disappears.

Many of the katavothra being insufficient to give

passage to all the water in the rainy season, a temporary lake is formed around the mouth of the chasm, which then becomes still further obstructed by pebbles, sand, and red mud, thrown down through the turbid waters. The lake being thus raised, its waters generally escape through other openings, at higher levels, around the borders of the plain constituting the bottom of the enclosed basin. In some places, as at Kavaros and Tripolitza, *where the principal discharge* is by a *gulf*, in the *middle* of the plain, nothing can be seen over the opening in summer, when the lake dries up, but a deposit of red mud, cracked in all directions. But the katavothra is more commonly situated at the foot of the surrounding escarpment of limestone; and, in that case, there is sometimes room enough to allow a person to enter in summer, and even to penetrate far into the interior. Within is seen a suite of chambers communicating with each other by narrow passages, and M. Virlet relates that in one instance he observed, near the entrance, human bones embedded in recent mud, mingled with the remains of plants and animals of species now inhabiting the Morea. 'It is not wonderful,' he says, 'that the bones of man should be met with in such receptacles, for, so murderous have been the late wars in Greece, that skeletons are often seen lying exposed on the surface of the country. In summer, when no water is flowing into the katavothra, its mouth, half closed up with red mud, is marked by a vigorous vegetation, which is cherished by the moisture of the place. It is then the favourite hiding-place and den of foxes and jackals; so that the same cavity serves at one season of the year as the habitation of carnivorous beasts, and at another as the channel of an engulphed river.

'Near the mouth of one chasm Mr. Babbage and his companions saw the carcase of a horse in part devoured, the size of which seemed to have prevented the jackals from dragging it in. The marks of their teeth were observed on the bones, and it was evident that the floods of the ensuing winter would wash in whatsoever might remain of the skeleton. It has been stated, that the waters of all these torrents of the Morea are turbid where they are engulphed, but when they come out again they are perfectly clear and limpid, being only charged with a small quantity of calcareous sand. The points of efflux are usually near the sea-shores of the Morea, but sometimes they are submarine; and, when this is the case, the sands are seen to boil up for a considerable space on the surface of the sea, in calm weather, in large convex waves.'

Readers will excuse this long extract, since it bears so much on the question, more especially as in this chapter I propose to do little more than quote instances of caves described by others. I need not dilate further upon caves which are formed where rivers enter, though the question of the deposit of bones may require more consideration. I would just, however, draw attention to the portion of this quotation which I have marked in *italics*.

It will be remembered that in a former chapter I described a large enclosed valley, called the Swede's Flat, in this district. Thus it was mentioned that the natural shape of the flat ought to make it a lake, but that whatever water was received by it *ran underground, either at the sides or middle*, and, where it goes at the sides, *cavities and hollows are seen under the limestone*, which crops out much water-worn and honeycombed. There are also hollows high up on the sides, and on the islands previously described. These serve as chains for the water when either of those below are stopped, or when they cannot carry away the water as quickly as it comes. The drains in the middle of the flat are those where by far the greatest quantity of water flows away. These are deep circular depressions, covered in summer with *caked mud*, cracked in every direction, and mixed with a great deal of sand.*

There are many other places in this district where large swamps, when overflowing, let the

---

* Might not the enclosed valleys of the Morea be the remains of chalk atolls? This interesting question is worthy of attention. The resemblance of the Swede's Flat to the Greek valleys is very great, as far as a description will enable me to judge of the latter. However, I am more and more convinced of the probability of the former being an upraised atoll. I was enabled, a short time since, to inspect a section of sixty feet of the bottom of the flat, where a well had been recently sunk. The deposits were just such as are described to be peculiar to the lagoons of coral islands. The bottom strata were much honeycombed, and through the crevices there was a *rush of wind* which extinguished the lights.

surplus water flow under the limestone, and these localities, which are caves in course of formation, are hollow passages, which can be followed for some distance, and much honeycombed by the passage of the water. I wish to direct my readers' attention to these facts, because they will assist much in explaining phenomena that will be mentioned hereafter; but I would have it, above all, remembered, that sometimes, in consequence of the stoppage of the drainage at low levels, the entrance to caves forming passages to subterranean streams may be found *much above the ordinary level of the country*.

And now for the osseous deposits. Is it necessary, it may be asked, that rivers, or underground streams, supposing the caves to be formed by such, should always bring down bones? It must be remembered, in answering this, that other things besides bones are found in caves where the deposit is recent enough to have them undecomposed. However, when we bear in mind that of all this débris, borne down by an ordinary stream, the bones of animals are the only things calculated to resist the action of decomposition, it is not astonishing that nothing else is found after a long course of ages. That other substances have been carried into the caverns, and subsequently decomposed, there can be no doubt. There are in every case, in addition to the stalagmite, deposits of fine black dust, or else, if the moisture is in excess, a finely-levigated black mud, such as is, under ordinary circumstances, derived from the decay of carbonaceous matter. We must consider the bone stalagmite, then, not as deposited in the manner in which it is found, but as mingled in the first instance with the ordinary débris of a rapid stream passing over a locality which was ordinarily dry land, and not the bed of a stream.

In corroboration of the view that streams would not, after a long period, leave any record except bones, we may cite the instances of those beds of former streams, revealed to the geologist by the upheaval of the land. The Pampæan formation is a case in point. This has been celebrated for enclosing innumerable bones of immense animals belonging to a former period of the earth's existence, including the Megatherium, the Megalonyx, Mylodon, Macrauchenia, Toxodon, &c. This large formation, which extends over many thousands of square miles, is composed of a red or brownish ochreous mud, and is now proved to have been the former bed or estuary of the Rio de la Plata. Now, though this large river must have conveyed down many other substances besides bones, these are the only, or nearly the only, things preserved.

Again, the upheaval of the land in many portions of the Australian continent shows, as banks of rivers, what has formerly been the bed. These seldom or never contain any drift wood or vegetable matter, but bones of animals have been occasionally found in them. And, indeed, a moment's reflection shows how this happens. Wood and light particles would float for a long time, and be carried out to sea, whereas animals drowned, or otherwise carried down, would float for a time, and then finally sink, and be buried in the mud. Now, if in the beds of rivers where the drowning or carrying down of animals is rather the exception than the rule, bones are found, how much more so in the beds of rivers which have emptied themselves into caves! For these would never take their course but in times of flood, when the waters invade the land, and drown many land animals; and if, as I shall show in the case of Australia, the most predominant bones are those of animals which burrow underground, and thereby the more liable to be drowned in sudden floods, there will be no difficulty in accounting for the osseous deposit in caverns.

In enumerating the different kinds of caves at the commencement of this chapter, I spoke of two other kinds, namely, beach, or those which have served as dens for wild beasts, and caves which serve as places for the egress. Of the former, the celebrated Kirkdale Cave is a good example. It was found by accident in 1821, in quarrying stone in the limestone peculiar to that part of Yorkshire. It was a long narrow passage, twenty-four feet long, and so low as to prevent a person walking upright. The floor was of stalagmite, but underneath was a bed of mud containing many bones. 'The surface of the sediment, when the cave was first opened, was smooth and level, except in those parts where its regularity had been broken by the accumulation of stalagmite, or ruffled by the dripping of water; its substance was an argillaceous and slightly micaceous loam, composed of such minute particles as could easily be suspended in muddy water, and mixed with much calcareous matter. That seems to have been derived, in part, from the dripping of the roof, and, in part, from comminuted bones.'*

There was a great variety of bones of different animals found in the mud we have described; but from the fact that hyæna teeth and bones were more numerous than any other, and that the bones of other animals were broken and quarried, besides the great quantity of hyæna dung mixed up in the mud, no difficulty was found in concluding that this cave must have been a den for hyænas, and that the bones were those of the tenants, mingled with those of their prey, which they had dragged thither to devour.

With this description of cave we have very little to do in this work, as Australia, with the exception of, perhaps, one lion, possesses no predaceous animal, unless the dingo be considered one; and this does not live in dens, or, at any rate, is glad to eat his prey wherever he can find it.†

---

* Buckland's *Reliquiæ Diluvianæ*.

† I will insert, however, a quotation from a letter of the Government geologist of Victoria, read before the Geological Society, London, June 1, 1859:—' The only other interesting discovery of the survey is the bone-cave at Gisborne, about twenty-five miles north of Melbourne. . . . . In it, embedded in light, powdery, and perfectly dry soil, we found great quantities of the osseous remains of birds and mammals, the most remarkable being perfect skulls of the dingo, the Tasmanian devil, and another carnivorous animal, which M'Coy thinks is quite a new genus. The skull is in shape somewhat similar to that of a

We have, however, to give an instance of sea-beach caves, and for this purpose cite the note of Sir H. De La Beche's 'Geological Observer,' where he speaks of the Paviland Cave, Glamorganshire, and of the human remains found therein:—'The cave in which these remains were discovered is one of two on the coast between Oxwick Bay and the Worm's Head, part of the district known as Gower, on the west of Swansea, and formed, in great part, by carboniferous or mountain limestone. It is known as the Goat's Hole, and is accessible only at low water, except the face of a nearly precipitous cliff rising to the height of about 100 feet above the sea. The floor at the mouth of the cave is about thirty or forty feet above high-water mark, so that, during heavy gales on shore, the spray of the breakers dashes into it. Beneath a shallow covering, Dr. Buckland discovered the nearly entire left side of a female skeleton. Close to that part of the thigh-bone where the pocket is usually worn, he found laid together, and surrounded by rubble, about two handfuls of small shells of the *Nerita littoralis*, in a state of complete decay, and falling to dust on the slightest pressure. At another part of the skeleton, viz., in contact with

domestic cat, but not more than half the size, and there are only two molars. The roof and sides of the passage were narrow, and were quite smoothed and polished, evidently from the frequent passage of the animals that have inhabited the cave. When discovered, all these passages were so completely filled up with earthy matter that no animal much larger than a rat could have obtained entrance. When cleaned out, some of them were four feet high.'

the ribs, he found forty or fifty fragments of small ivory rods,* nearly cylindrical, and varying in diameter from a quarter to three quarters of an inch, and from one to four inches in length. Their external surface was smooth in a few which were least decayed, but the greater number had undergone decomposition. Fragments of ivory rings were also observed, supposed, when complete, to have been four or five inches in diameter.

'Portions of elephant tusks were obtained, one nearly two feet long, and Dr. Buckland inferred that the rods and the rings had been made of the fossil ivory, the search for which had caused marked disturbance of the ossiferous ground, the ivory being then in a sufficiently hard and rough state to be worked. Charcoal and pieces of nine recent bones of sheep, oxen, and pigs, apparently the remains of food, showed the cave had been used by man.

'The toe-bone of a wolf was shaped, and it was inferred, that it had been probably employed as a skewer. As regards the date when this cave may have thus been worked for its ivory, and the woman buried, Dr. Buckland calls attention to the remains of a Roman camp on the hill immediately above the cave. Amid the disturbed ossiferous ground there were not only recent bones, but also the remains of edible *Buccinum undatum* (whelk), *Littorina*

---

* Similar rods of ivory were found by Sir Christopher Wren in sinking for the foundations of St. Paul's Cathedral, London. The place was supposed to have been an old Roman cemetery. Underneath were found sand and eocene shells (London clay).

littorea (periwinkle), &c.' I have already stated that this quotation was more for the sake of illustrating the kind of cave meant than for any direct reference it has to Australia. It is, however, singular, that in all North Australia caves have been discovered which have evidently formerly been tenanted by the aborigines. The walls around are covered with rude frescoes in red ochre, containing emblems as curious for their great antiquity as showing some remote connection with Hindoo designs.

With reference to the egress caves, or passages which gave egress, whose source is not known, not so much is to be said. There are none in Australia, nor, indeed, are there many in the whole world. They are not ossiferous. The origin of the water in them is not known, but several theories are extant on the subject. One is, that they are connected with immense reservoirs of water, which collect from infiltration, like artesian springs. This is very probable. Most of my readers are acquainted with Humboldt's description of one he visited in South America, near the convent of Caripe. As this is a good instance, and the account is replete with interest, its insertion here will be excused, in a condensed form, of the account from the 'Personal Narrative' which refers to it.

'The Cueva del Guacharo, as preserved in the vertical profile of a rock.—The entrance is towards the south, and forms an arch eighty feet broad and seventy-two high. The rock which surrounds the grotto is covered with gigantic trees; but this luxury of vegetation embellishes not only the external arch, it appears even in the vestibule of the grotto. We saw with astonishment plantain-leaved heliconias eighteen feet high, the maya palm-tree, and arborescent arums, following the course of the river, even to those subterranean places. The vegetation does not disappear till about thirty or forty paces from the entrance. We measured the way by means of a cord, and we went on about 430 feet, without being obliged to light our torches. Daylight penetrates far into this region, because the grotto forms but one single channel, keeping the same direction from southeast to north-west. When the light began to fail, we heard from afar sounds of the nocturnal birds. As we advanced into the cavern, we followed the banks of a small river which issues from it, and is from twenty-eight to thirty feet wide. We walked on the banks as far as the hills, formed of calcareous incrustations, permitted us. Where the current winds among very high masses of stalactites, we were often obliged to descend into its bed, which is only two feet deep. We learned with surprise that this rivulet is the origin of the river Caripe, which, at the distance of a few leagues, is navigable for canoes. The grotto preserves the same direction, breadth, and height for 1,458 feet. We had great difficulty in persuading the Indians to advance as far as a spot where the soil rises abruptly at an inclination of sixty degrees, where the torrent forms a small subterranean cascade. We climbed this, not without difficulty. We saw that the grotto was perceptibly contracted, retaining only forty feet in height, and that it continued stretching to the north-east without deviating from its original direction, which is parallel to the valley of Caripe.'

The illustrious traveller then goes on to consider the subject of caves generally, which he treats in a manner worthy of his patient acuteness; but his opinions are rather behind the present state of science.

Other instances might be mentioned of rivers issuing from caverns, and causing the same characteristic appearance of a straight narrow channel, of nearly equal width and height, different entirely from those which have been formed by floods, by the absence of tortuous windings, wide chasms, and deep fissures.

There was a river issuing from a cave, precisely similar to that of Caripe, near Tehnilotepec, in the western Cordilleras of Mexico; but in the night of the 16th of April, 1802, the river suddenly ceased flowing, bringing great ruin on the inhabitants of the countries through which it formerly ran; probably, some subterranean disturbance connected the reservoir with another outlet, or turned it into a lower stratum.

We have now gone through the description of caves, according to the classes into which, for convenience, I have divided them. I have not mentioned many that are probably more interesting than those I have described, because I only wished, in this chapter, to illustrate certain principles and theories, and, accordingly, only cited those which were most apt for the purpose.

I might, for instance, have described the Mammoth Caves, in Kentucky and Tennessee, which are certainly the most remarkable in the world. Many of them have been descended for hundreds of feet, and streams of water have generally been found in them. Some of them have been followed for many miles; indeed, so common a feature is this of the country, that they cease to attract attention. They are generally like other caves, whose roofs and sides of limestone are encrusted all over with stalactite. There is one cave, in the Cumberland mountain, of such great depth that its bottom has not been reached.

The mention of this district, where the rock is all limestone, and of so loose a texture as to be easily undermined with caves, reminds one of the district of which I am now treating. Here the limestone is loose, and covers immense tracts of country, and, consequently, caves are so numerous as to be scarcely a matter of comment. In their description the next two chapters will be occupied, and it is in order to understand the import of the various appearances, that I have dealt generally with the subject in this chapter.

I have shown it to be, commonly, that the theory

for the osseous deposits must vary in every case. What views on the subject will be required for the osseous breccia in the cover of this district, will be the subject of the next chapter.

CHAPTER XII.

CAVES.

CAVES IN GENERAL.—CAVES AT MOSQUITO PLAINS.—FIRST CAVE.—SECOND CAVE.—THIRD CAVE.—DRIED CORPSE OF A NATIVE.—ROBERTSON'S PARLOUR.—CONNECTION BETWEEN IT AND DEEPER CAVES. — CORALLINE LIMESTONE. — BONES. — BONES OF RODENTS.—OTHER BONES.—MANNER IN WHICH THE CAVES WERE FORMED.—FORMER LAKE NOW DRAINED BY A CREEK.—EVIDENCE OF FLOODS.—NO EVIDENCE OF THE DELUGE.—CONCLUSION.

OF all the natural curiosities a country can possess, none tend so much to render it famous as the existence of large caves. There is such an air of mystery in the idea of long subterranean passages and gloomy galleries shut out from light and life—so little is known of their origin, and they are generally accompanied with such beautiful embellishments of Nature—that one is never tired of seeing them, or of hearing the description of those that cannot be visited. Thus, every one who may otherwise never have heard of Adelsberg, has heard of the Adelsberg caves, with the renowned pure white stalactite, which, hanging from the roof like an immense snowy curtain, is so translucent as to show torches placed on the inner side. In like manner, every one has heard of the caves in the

Peak of Derbyshire, where visitors are carried in a boat, by a subterraneous river, along a passage scarcely two feet high, before they can inspect the inner portion. Persons who have never read Humboldt's 'Personal Narrative' have at least heard of the Guacharo caverns, in South America, described in the last chapter, which are tenanted by thousands of owls, whose screeching makes the place like a den infernal. Few are, perhaps, aware of the existence of the caves in New South Wales, described by Sir Thomas Mitchell, and fewer still know of those in Tasmania.

But, wherever such natural curiosities are known, they do not fail to give great importance to the place, making it as noted as if it possessed a burning volcano or a geyser spring. I am not aware that any attempt has been made to describe the caves we possess in South Australia. Some occasional tourist may have notified, in a stray newspaper paragraph, the fact that such things existed; but, as far as giving an account of their rich and varied beauties, as far as relating the extraordinary natural curiosities that are to be met with in them, nothing at all has been done. And yet in point of magnitude, in point of splendour, and in a scientific view, they do not yield in importance to any of the wonderful phenomena enumerated above. In this chapter I propose to give an account of them, which, to do them justice, must be rather lengthy; for to bring the description within small limits would cause many things which are of scientific importance to be omitted. If the narration is long, the presumed interest of the subject must be the apology.

About twenty-five miles north of Penola, on the sheep-run of Mr. Robertson, in the midst of a swampy sandy country, plentifully covered with stringy bark, a series of caves are found, whose internal beauty is at strange variance with the wildness of the scenery around. There is nothing, outwardly, to show that any great subterraneous excavation might be expected. The entrance to them is merely a round hole, situated on the top of a hill; and, were it not for the existence of certain temporary huts, and other unmistakable signs of the former frequent visits of Bush excursionists, one might be inclined to pass the place without noticing anything peculiar.

On going to the edge of the hole, a small sloping path is observed, which leads under a shelf of rock, and, on descending this for a depth of about twenty-five feet, then it is one gets the first glimpse of the magnificence enshrined below. The observer finds himself at the entrance of a large oblong square chamber, low, but perfectly lighted by an aperture at the opposite end, and all around, above and below, the eye is bewildered by a profusion of ornaments and decoration of Nature's own devising. It is like an immense Gothic cathedral, and the numbers of half-finished stalagmites, which rise from the ground like kneeling or prostrate forms, seem worshippers in that silent and solemn place.

The walls are pretty equal in outline, generally unbroken nearly to the floor, and then, for the most part, they shelve in as far as the eye can reach, leaving a wedge-shaped aperture nearly all round. This seems devised by Nature to add to the embellishment of the place; for in the space thus left, droppings of limestone have formed the most fanciful tracery, where pillars of every shape wind into small groups, like garlands of flowers, or stand out like the portico of a Grecian temple, the supports becoming smaller and smaller till they join like a mass of carved marble.

At the farther end there is an immense stalactite, which appears like a support to the whole roof. This shuts from the view the aperture in the roof behind it, so that the light steals in with a subdued radiance, which mellows and softens the aspect of the whole chamber. The pillar is about ten feet in diameter, and, being formed of the dripping of limestone from above, in successive layers, seems as though it owed its elaborate appearance to the hand of Art, not the least beautiful part of it being that it is tinted by almost every variety of colour, one side being a delicate azure, with passages of blue and green and pink intermingled; and again it is snowy white, finally merging into a golden yellow. It stands upon a raised platform of stalagmite, which extends some way down the chamber, about three feet high, at the end of which is the pillar.

This platform has been a mass of small stalag-

mites, which are now joined together by successive droppings, that have covered them over in a manner not unlike the spreading of a linen cloth. At the south end (the entrance), the cave looks as if prolonged behind each side of the narrow opening. But this is not the case. There is merely the same continuance of columns, like those found all round; somewhat larger, indeed, and joined together so closely as to make the spaces look like the pointed arches of a mediæval crypt. The whole length of the cavern, as near as I could ascertain, is about 190 feet, the width about forty-five feet, and the height twenty feet. The floor is deeper towards the middle, so that the latter measurement varies. Its length would be much greater, and it would run into the next cave, but that it is blocked up by the large stalactite I have described above.

On going round this, and observing, still on every side, the stalactite pillars, the opening which lets in the light to the north end is seen. There was evidently no aperture here formerly, as a pile of broken limestone shows the roof to have fallen in; and, by the manner in which the damp has rounded the sharp fragmentary outlines, by the way the heap is covered by creeping plants, it declares itself to have happened a long time since. This inlet is larger than the one at the entrance.

The second cave that now meets the view is different in many particulars from the former. It is smaller, and so thickly studded with stalactites as to render a clear glance through it impossible. These are not like those of the former cavern, differing inasmuch as they are all very white, and mostly broader at the top than at the base, giving them the appearance of groined arches. Some are thin, and look, from the manner in which they are deposited, as if they were gracefully festooned in honour of some festival; some are mere delicate shafts, and every now and then some large unfinished stalagmite appears in the form of a veiled statue, mysteriously enshrouded in heavy white drapery.

When this chamber has been nearly traversed, on looking back, it is surprising what a different aspect it bears; one would think a dense avenue of statuary before some palace had been passed, so solemn, so great, and yet so life-like are the curious wreathed and twisted columns, with their numerous groupings and strange varieties of form. At the end of this cave (it is not half the length of the first) there is another aperture open to the light, caused also by the falling in of the rock, which once arched it over. It is a large circular hole, whose sides are precipitous, with a smaller pile of broken stone in the middle, as in the one last mentioned. It was here that many years ago some natives destroyed 300 sheep, by throwing them from above on the hard rock below. This was about the time they were committing many outrages, including the murder of Mr. Brown. How the settlers revenged themselves is shown by something farther in the cave, which will be presently noticed. This opening is the last through which light gains admission to the vaults, and the entrance to the last cave is on one side, in a line with that just quitted.

This one is so thickly studded with stalactites, and these, sometimes, so very wide at the base, that from the outside it seems like a carefully-arranged scene, which, from the interminable variety of form, or magic effect of light and shade, might easily be thought intended to represent a fairy palace. On proceeding a little way, the ground becomes painfully uneven. You have to climb over boulders, whose summits almost reach the roof, or you have to descend into what might almost be called pits, the more rough and uneven because of their natural ornaments.

Very soon the cavern becomes as dark as night, so that no further exploration can be made without candles, and, even with these, the utmost caution is necessary, as there are pits, caverns, and holes in all directions, some of them leading to other small subterranean passages. There is one, in particular, which is a great fissure, extending nearly from side to side. It is very deep. The sides are smooth and slippery, and, as light is thrown into its gloomy depths, the sides are seen to be divided in some places into columns and pillars, making even that dark place elaborate with natural architecture.

Farther into the cave the roof becomes lower and lower still, surmounted with the ghostly white stalactites, and, at last, the passage onward is so small that one must stoop very low in order to proceed.

It is not without a shudder that one goes through this passage. Far away from the light of day, this groping along a small vault makes one dread to be bent down between stone walls, unable to stand straight or breathe freely. The passage widens, however, when the last chamber is reached. There are few stalactites here, but the number of boulders increases, so that to explore the place is to climb and scramble from rock to rock. At the upper end there is an immense mass of stone, by scaling which the cave is seen to narrow, so that human beings can hardly go farther. There are, however, many passages at either side of this and the other chamber, some of which have been explored, and it would appear that they are continuous to an immense depth underground. This, therefore, may be called the last chamber, though filled to bewilderment with fissures and galleries which may lead into as many more.

A painful stillness reigns in this last cavern, which becomes positively unbearable, after remaining a little time. Humboldt, in his account of the caves of Guacharo, complains that the noise of the birds dwelling there gives an awful addition to the horror of those underground vaults; but any noise would be less dreary than the dead silence which reigns here. Whether it is that the air is hot and close, or whether the depth compresses the atmosphere beyond its usual density, I cannot say, but certainly the quiet presses painfully upon the sense of hearing, and the closeness gives a feeling of smothering which adds to the horror of a place deep in the earth and far from the light of heaven.

At the side of one of the boulders, on the right-hand side in entering, in a crevice between it and the wall where Nature seems to have made a natural couch, lies, in the position of one asleep, with the head resting on the hand and the other limbs reclining, the dried and shrivelled corpse of a native, but slightly decayed, and almost petrified by the droppings of the limestone. It is known to have been there for many years without decomposition, though the fingers and feet became annually more encrusted with stalactite.

The history of his coming there is a sad one. The blacks, in addition to the destruction of the sheep spoken of above, committed murder and so many acts of violence that the settlers resolved to be avenged. They assembled, and set out with the significant motto, 'Let not your right hand know what your left hand doeth.' The natives resisted desperately; some were shot in every part of the country. One, wandering near these caves, was seen, and brought to the ground by a rifle-ball. Badly wounded, he managed to crawl away unobserved, and, thinking that he would be sought for as long as life was in him, crept down into the lowest and darkest recess of the cavern, where he rightly judged few would venture to follow. There he lay down and died. Time went on. Not a tear was shed over him as he lay there uncoffined, but drops of water fell upon him from the rock above; and when, a long time after, his remains were discovered, the limestone had encased him in a stony shroud, which to this day preserves his remains from decay.*

The limestone alone will not, however, explain the absence of eremacausis. The peculiarity of the atmosphere has something to do with it. I noticed, near the entrance of the last cavity, the body of a sheep, which had evidently fallen from above while the animal was too incautiously browsing on the tempting foliage. It had been there some time, yet the flesh seemed as if but lately killed. The chemical property of the air does not materially differ from that above, and no satisfactory reason appears why the chemical constituents should not, once the vital stimulus has ceased, re-act upon themselves in this case as in every other. The same thing, however, is observed in many vaults, and probably the uniformity of temperature bears a part in the phenomenon of which the renowned kings of Cologne and the mummies of the Italian cemeteries are instances.

On leaving this last and lonely chamber to return to the light, a narrow fissure, richly wreathed with limestone, is observable on the right hand

---

* An enterprising showman has since stolen this body. It was once recovered from his hands, but was finally carried off. The whole history of the larceny, and the attempts of the Government to recover the body, form a very amusing incident in colonial history.

going out. Proceeding a little way down, a large vaulted chamber is reached, so perfectly dark and obscure that even torches can do but faint justice to its beauty. Here, above all other portions of the caves, has Nature been prodigal of the fantastic ornament with which the whole place abounds. There are pillars so finely formed and covered with such delicate trelliswork, there are droppings of lime making such scrollwork, that the eye is bewildered with the extent and variety of the adornment: it is like a palace of ice, with frozen cascades and fountains all around. At one side, there is a stalactite like a huge candle that has guttered down at the side; at another, there is a group of pillars, which were originally like a series of hour-glasses, set one upon another from the roof to the ground, and the parts bulging out are connected by droppings like icicles, making them appear most elaborately carved. In addition to this, there is above and below—so that the roof glistens, and the ground crackles as you walk—a multitude of small stalactites, which fill the whole scene with frostings that sparkle like gems in the torchlight. In one of the passages leading away from this chamber there is an opening, which, after being followed for some distance (on all fours, for it is exceedingly small), leads into another spacious chamber, full of stalactites, open to the sky at one end by a wide aperture. This latter cave was known for a long time by the name of the Deep Cave, and was thought to be quite disconnected with the ones just described. Indeed, it was at one time believed to be almost inaccessible, as there is a clear descent of about thirty feet from the roof to the floor of the cavern, but quite lately there was a communication found between the two. There is nothing peculiar in this chamber making it differ much from the last. Of course the festooning of stalactites is as fanciful and full of beauty here as elsewhere, except that they are rather less numerous, and there is a little less light to view them by. At the side of this cave there is another cave, probably also communicating; the passage has been discovered at the same time. This is exceedingly deep, probably over sixty feet, and only a wide spacious chamber. As there is no possibility of descent except by a rope, and as I was informed that the cave possesses little that is interesting, I preferred to wait for its exploration until a more practicable passage should be found between it and its neighbours.

This is the last of the subterranean beauties, and, on emerging towards the opening, the fresh air and more luminous aspect come gratefully upon the senses. Amazed and stupified as you may be with the beauties left behind, one feels, as the eyes become dazzled by the approaching light, that the greatest beauties of the earth lose half their charms when shut out from the heavenly radiance of the sky.

I have now to allude to some organic remains and other curiosities found in the caverns. On one side of the first chamber of the cave just described there is a fine section of the coralline fossiliferous limestone of which the rock is composed. Here are seen immense masses of the *Cellepora gambierensis*, which is the predominant fossil of the formation. It is standing upright, shrub-like, and much branched, exactly in the position in which it grew. This must have been very near the main reef, or perhaps formed part of it; at any rate, it has not been disturbed since its growth, and must descend to a much greater depth than the floor of these caves. Shells are common on the rocks, especially the *Pecten coarctatus*, which has been so often spoken of in a former chapter of this work, and at a small distance nearer the entrance the coral entirely disappears, and white limestone is found in layers varying in thickness from one to six feet.

Next among the important organic remains of the cave are the bones. It has not been mentioned, in treating of osseous caves, that the bones of animals when found in caves, if like existing species, were always much larger than any which are contemporaries with man.

In Germany, in Italy, and in many other places wherever bones were searched for, they were found, more or less abundantly, in every case, similar to animals at present existing, but of a much smaller size. This latter point is of much importance, and may be stated as having become almost a law in geology, as it is applicable to almost every instance known, that the animals immediately preceding those at present existing on the earth were identical in every particular with the present, only very much larger.

Knowing these facts, and also knowing that our caverns were as ancient, according to appearance, as any mentioned above, there is nothing surprising in finding osseous deposits in them also. Long before I had visited these caves, my attention was called to what was stated to be a small pile of bones, which were found one day by the accidental breaking of the stalagmite with which they were covered over. On examining the spot indicated, I found they were in the raised platform, at the foot of the large stalactite, in the first cave alluded to above.

This platform is about fourteen feet long by eight broad, and I have no hesitation in saying that, excepting the thin layer of stalagmite on the top, it consists nearly entirely of bones. Nor is this all. During the whole length of all the caves, wherever the floor is sufficiently level to enable one to perceive it, there is a constant reappearance of the broken bones, whenever the limestone pavement is broken through. How deep the deposit goes, I do not know, but in the platform just named I was able to scrape away almost to the depth of two feet, and found the deposit as thick as ever.

The extraordinary manner in which they are agglutinated together is also worthy of remark. They are not found in any regular position, such as would be imagined had their owners lived and died where their remains now lay. Heads, jaw-bones, teeth, ribs, and femurs are all jumbled and concreted together without reference to parts. The quantity of small animals it must have taken to form a deep deposit of their bones—perhaps two feet deep, ten wide, and of indeterminate length—must have been something prodigious, for they are compressed into the smallest possible space, and must have decomposed from exposure. How they came there—a question which has puzzled all geologists—I will allude to by and by. We have first to examine to what animals they belonged. The bones which most predominate are evidently those of some animal belonging to the order of Rodents. The skulls, teeth, and bones of these abound, perhaps in the proportion of three to one of any other description, and, though numerous, it was with considerable difficulty I could find one entire skull. It may be described as a low flat head, with the incisors of the upper jaw coming abruptly out at a curve from the bony palate, the orbits large, with the molars on each side pointing outwards. The incisors of the lower jaw do not meet those of the upper when both are *in situ*, and there is a considerable hollow between the three molars and the lower part of the incisors. There are sixteen teeth in all—four incisors, and on either side of both upper and lower jaw, three molars. In this case, as indeed in all the Rodentia, there is a great distance between the incisors and the back teeth, but, as it appeared to me, greater in the skulls I am now considering.

At first, I was rather at a loss to make out the

exact species to which the remains formerly belonged. The size (about an inch and a quarter long,

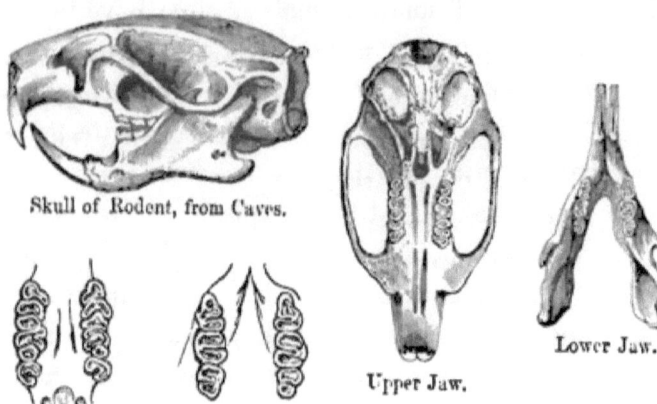

Skull of Rodent, from Caves.

Teeth of Upper Jaw, enlarged.   Teeth of Lower Jaw, enlarged.

Upper Jaw.

Lower Jaw.

and three quarters of an inch wide) made me inclined to refer them to the jerboa, described by Sir T. Mitchell as occurring on the Murrumbidgee; but I looked in vain for the long tibia which should be in the neighbourhood of the skull of such an animal. Besides, the teeth were only three in number, and, though it is suspected that the fourth tooth disappears from the adult jerboa, their structure was against such a conclusion. In the latter animal the enamelled edge makes a sort of sinuous or waved edge around the whole tooth; but in the ones under consideration there were three distinct septa in the enamel of the grinding surface on the first tooth (the anterior and largest), and two in the two others. After having referred these teeth to an animal very closely allied to our domestic mouse, only much larger, which I was led to do after some consideration, I concluded that they belonged to an extinct species, and confirmed the law as to size which has just been alluded to. I have since found, however, that in this I have been mistaken. My attention was often called to little mounds of sand in the plains, where rushes grew abundantly, and these were bored on every side by small burrows. For a long time I was under the impression that these were caused by bandicoots (*Perameles*), which burrow underground for the roots. One day I caught one of the little brown creatures, which I constantly saw running from hillock to hillock, and into their burrows. To my astonishment, I saw that the teeth corresponded in all particulars with those of the rodents in the cave. As the species is, to the best of my belief, new, I will here describe it. It is of a dark-brown colour, the fur thin and fine, filled with longer hairs of a lighter colour. The anterior limbs have four complete toes, which are sharp and compressed. There was no rudimentary thumb; the hind feet have five toes, which have also sharp, compressed nails. The two external ones are much shorter than the others; the muzzle is short and blunt, and the teeth are more similar to the true or common rat than any of the Rodentia.

Their dentition more nearly resembles the true rat tribe than any of the same family; the whole animal resembles the Cape otomys (whose dentition is also an approach to the true rat), and doubtless will be found to form a link of connection between the two species. I must further remark, that it does not belong to any of the present catalogued species of rodents belonging to Australia. There are fourteen species peculiar to Australia, and two water rats. Two are peculiar to Tasmania, two to Port Essington, and the rest common to the southern part of the continent. As far as I am aware, it is not one of these. However, its bones are the predominant ones in the cave, and the habits of the animal easily explain the peculiarity.

As it will be shown hereafter, these caves owed their origin to times of flood or inundation; and those animals whose habits led them to burrow in low flat lands would be, of all others, the most likely to become the first victims of such a visitation. The foregoing drawings of the skull are from a cave specimen. The conclusion is rather ludicrous.

'Parturiunt montes, nascetur ridiculus mus.'

These bones, instead of belonging to extinct animals, are those of animals existing within a short distance of the caves. The same may be said of the bones by which these are accompanied.

The first and most common, next to those above, are long jaw-bones, with four molars, three false molars, one canine, and three incisors on each side; the condyle a flat well-defined hinge, and the coronoid process sloping back at a very obtuse angle, so as not to be raised much above the plane of the jaw. These features would seem to imply an animal with a long, low, flat head, of predatory habits, bearing great resemblance to the long-nosed bandicoot (*Perameles nasuta*), to which, or to a nearly allied species, no doubt the bones belonged. These animals also burrow in low grounds.

The second are the jaws of an animal not unlike the *Myrmecobius*, with two false molars more than the native cat, and the condyle very imperfectly developed. I must mention that the angular process or inflection of the side of the jaw was most perfect in this instance, making it extremely doubtful whether the animal was of the marsupial order; yet the animal was of that order, and the species is yet existing in the neighbourhood as the *Phascogale penicillata*, or native squirrel, a pretty little animal, eight inches long, with a long pencillated black tail, and the rest of the fur a light grey, exquisitely soft and delicate. This little animal is most destructive and pugnacious, living in dead hollow trees, and I have only seen it near low land.

To this family also belong the bones of a small animal not uncommon in the stalagmite. The jawbones are about five-eighths of an inch long, and distinguished by the extraordinary sharpness of the needle-like protuberances on the crown of the molars. The animal previously mentioned has very pointed crowns to the molars, and false molar teeth, but those of the latter are quite as minute and sharp as those of the bat, to which animal the dentition bears a strong resemblance. I presume the animal possessing them was the *Phascogale pygmæa*, a small variety of the *Phascogale*, which is not now common. It is mostly found in the more northern parts of this district, and frequents trees, burrowing near their roots.

The next was an animal possessing canine teeth, which bore an extraordinary disproportion to the others. There were in addition, on each side, five molars, one false molar, and three incisors. The condyle, coronoid, and angular process much resembled those next-mentioned animals, probably both insectivorous and carnivorous, from the form of the teeth. Next were the bones of an animal as nearly as possible resembling our native cat (*Dasyurus Maugii*, or the spotted opossum of the early settlers), though not identical. I could find no perfect adult specimen of the lower jaw.

Both these animals belonged to the same family of *Dasyurus*, but the first-mentioned, or smaller variety, with large canine teeth, does not at present exist, as far as my knowledge extends. The latter is a very common frequenter of houses in Australia, being as destructive and vicious as the rat at home, whose place in domestic economy it usurps in this colony. In its wild state it lives under rocks and stones, in fact, in any underground cavity, but it does not burrow, and only takes to trees when pursued, or at night in search of birds, which it kills while roosting. Besides these bones there were those of the vulpine phalanger, or common Australian opossum, and several others which are known to be common about the immediate neighbourhood.

It is to be remarked, however, that the bones

common are those of animals which burrow underground, and liable, from that cause, to be drowned by any sudden advent of water; also, the bones do not seem to be entirely deprived of gelatine, but they have the appearance of great antiquity. They are generally covered with a crust of lime, which easily scales off in thin plates, leaving the bone clean and perfect.

I will not now enter into a description of the other bones; it would take ages to classify them all, even were the difficulty less than it is, so I must content myself with stating that I could find no remains of a large animal, and it must have taken millions of individuals to raise the deposit that is formed. I may add, however, that the types of all the existing animals would not be much smaller. The kangaroo bones found in the Wellington Valley caves are at least three times the size of any now living; and the same may be said of the opossum. Those caves are, I understand, in a much older deposit, and probably the same may be said of the bones, though, from what I shall say in the next chapter, the large kangaroo may be still existing. Now, as to the way these bones came to be so congregated: had the mouse-bones been smaller, and near some Phœnician colony, we might suppose them to be relics of Pagan religious worship, for these people used to sacrifice mice in caverns, and make a tumulus of the bones. Such a theory would hardly do here. We must premise, first, that the animals did not live and die where they are found, for their remains are not associated with what we must expect, had they lived there, neither are their bones found in the state they would be in under such circumstances. Besides, the depth is too great, and the place too extensive, for any animal to live in as a place of shelter.

Some geologists are of opinion that most caves were formerly in the position of an underground current or river (not uncommon in limestone), which would carry down organic remains; but I can assert almost positively that there is no visible place for either the egress or ingress of water in these caves, unless by the roof, or through the meandering thread-like passages at the end. A river in the sense of a continued running stream there could not be, or even a creek, so that the theory will not meet the present case, so far as I have as yet seen.

Some, again, suppose the animals to have fallen from above; but though this would account for bones near the holes, it would not give a reason for a deep deposit extending the whole length of the passage. Some others agree that the bones could only have collected during an extensive inundation, which would cause them to accumulate, either by driving large numbers of animals into the caverns, or by the restless agitation of the waters above.

With this latter theory I agree, as the most consistent with observed facts. I have remarked before, that the caverns are on rising ground (another argument against a river). Now, suppose an inundation gradually covering the plains below, all living creatures (that were not drowned in the plains, which would not, as we have seen, be the largest number,) would take refuge on the hill. Let the waters still rise until a multitude of all the things that creep the earth are huddled on to the hillocks all around. Place a cave on the top; how rapidly would they take refuge therein, and as the swollen waters poured slowly into their last resource, what multitudes would leave their skeletons to mark the work of destruction, besides the floating bodies of those drowned by the first rush of the waters below, that would be carried down by the current or swept in by the wayward action of the fluid. This theory appears to me to be the most acceptable; and let us look, for one instant, at the curious corroboration afforded by the nature of the country around.

The caves, as I have said, are on the summit of a small hill, which is part of a low range running north and south. It is separated from another range on the west by a narrow flat, scarcely a quarter of a mile wide. This flat is singularly level, and where there are any eminences they have a rounded outline, making them look like islands on the flat. To the north this flat is closed, at about six miles from the caves, by a junction of the ranges. To the south, at about four miles, it opens out into a much wider flat, and then is closed by a junction of the ranges again, with the exception of a small opening, through which a rivulet or creek passes.

It will therefore be seen that, were it not for this opening, the whole flat would be a valley perfectly enclosed, and allowing no exit for the water, which would drain down from the surrounding hills. This would give rise to an inland lake, whose only drainage would be when the water was high enough to pour into the caves. In fact, the caves would be neither more nor less than the katavothra, or swallow-holes, of the enclosed valleys of the Morea, spoken of in the last chapter. Now, there is very good evidence that the creek which at present drains the flat has only been recently formed. When overflowing in winter, it enters very deeply into the banks, so that, in a few years, it will be much wider than it is now. As its greatest width is very small, there can be no question that its origin is very recent. Probably, as the range is rather lower here than elsewhere, its beginning was an overflow, when the inland valley was rather more full than usual. The flat itself, even if the existence of the caves were not known, would be ascribed to a lake, because the level appearance of the bottom and the nature of the sides are precisely similar to the Swede's Flat, already alluded to. However much the aspect of the country has altered since the occurrence of the water upon this lake, the appearance has not changed to such an extent as to leave the least doubt about the origin of the caves, when the ground is inspected.

With regard to the nature of these inundations, I do not think the country has been more liable to heavy rains than it is at present. Certainly a flood of water covering a flat, and converting it into a lake, would be an astonishing as well as a pleasing sight to the settlers here; but, if drainage were imperfect, the flat would be without the creek. I am sure one of our ordinary winters would produce all the results that are here evident, leaving the country around as little marked with ravages of excessive rain as it is at present,—and that is saying a great deal. I was once of opinion that there were here signs of an extraordinary inundation, but a more careful inspection has quite dissipated this notion. Floods there have been, and probably seasons of more violent rains than commonly seen at present, and perhaps of such violence as to be unexpected again, without some great change in our at present sleeping volcanoes.

It is singular that two phenomena should be accompanied with such similar results in countries so far apart as Australia and the Morea, yet there can be no doubt that our caves and their valleys, and our swallow-holes and katavothra, are in all respects identical.

The entrance to the caves at the Mosquito Plains is from above, and the shape of the descent into them, and the walls on each side, is exactly that of a watercourse. At the first descent, the stones have fallen down into a kind of slope directed towards the right side of the cave, which has a deep indentation, in consequence, just at the distance that the water would impinge upon it. Again, this hollow has a projection on the farther side, which has thrown the stream to the other side of the cave, where there is another indentation. From this the water has evidently been thrown off on to the big stalactite before described, at the foot of which all the bones have been deposited. It is easy to see that this stalactite is the only obstruction the water would meet in its course, and the occurrence of bones in any quantity here, and here only, is thus explained. I think that there is evidence here also to prove that these inundations took place many times, and that long periods of rest intervened, during which no water flowed at all. In the first place, the caves must have existed some time before stalactites were formed; and, secondly, those stalactites which reach from the roof to the ground would have been washed away, had the water been continually flowing.

Therefore, there must have been, first, the floods which scooped out the caves; and, secondly, the floods which piled up the bones at the foot of the stalactites formed during a period of rest.

For the first, a great many floods of water must have flowed to hollow out so large a series of caverns; probably every year, or nearly every year, during the summers or dry seasons of which the stalactites were forming. For the second, there must have been either one violent inundation, so as to drown all the animals in one great catastrophe (and of this there is no evidence), or there must have been successive quantities brought down by the annual flow in every winter season.

The eruption of Mounts Gambier and Shanck, and the volcanoes to the southward, may have caused very heavy torrents of rain and extraordinary floods, as these events generally do. Indeed, this must have been the cause of whatever other signs we see of floods here.

It may be remarked, that there must have been some sorts of holes or cracks in the limestone for the water to have flowed down in the first instance. But this is not necessary, for the mere infiltration of water through the soft and porous limestone, where it was exposed, would soon form a passage. But I think we may reasonably conjecture that the strata underground are full of cracks, apertures, and fissures. It has already been frequently stated, that the whole of the district from Mount Gambier to the Tatiara is composed of light limestone, formed of porous strata, which, though much disintegrated at deposition, would, in the course of time, settle down by its own weight, or become disintegrated by filtration. As it was all under the sea at one time, and as it was slowly raised from thence, each portion would be successively covered by shallow water exposed to the action of coast waves. This would break the corals and shells of the uppermost strata into fragments at first, and afterwards to an impalpable paste, which would harden into a very compact rock when dry, suffering entirely from the loose underlying shelly deposit. In the course of time, when the rock was quite raised from the sea, the most loose of the shelly parts would crack and loosen into fissures, leaving a space under the hard, concreted upper strata, thus giving rise to caves. In the district are many caves of which the hard roof never falls in to reveal their extent, and which are only known to exist by the hollow sound percussion of the surface gives, or by the boring of a well accidentally displaying them. In confirmation of these views as to their origin, I may here state what has been formerly mentioned, that, wherever the formation occurs, there are always about three feet of hard schisty limestone covering it. Secondly, caves are very common in the district; and, finally, I have seen the same thing in operation at Guichen Bay, where the loose shelly rock has been hardened by the mere action of the waves into a thick deposit above the proper formation, which remains loose.

Now as to stalactites. It was formerly stated, by many eminent chemists, that these could not easily be accounted for, as water would not dissolve carbonate of lime, or the ordinary limestone. It has, however, been since determined satisfactorily, that water will hold a certain quantity of carbonic acid in solution, and will then dissolve a certain quantity of lime. Water falling on grassy ground derives a quantity of carbonic acid from plants, and this, filtering through and evaporating, would leave the lime it had dissolved on the inner side as a little nodule, gradually enlarging by increasing deposition. Wherever the quantity of lime was small and pure, and the evaporation slow, crystallisation would take place, which is the case in nearly all the stalactites in these caves. I must mention that, with the exception of the ridge on which the caves are, there has been little or no upheaval, and no higher ground from which any stream might be derived for a long distance. The country around is singularly level and flat, destitute of anything like a large creek, or even of surface-water in a dry season. Devoid of rivers and hills, the aspect is far from pleasant for those whose tastes are with the poet, who said:—

'Rura mihi et rigui placeant in vallibus amnes.' VIRG.

Before concluding this description of the caves, there is one point which I am anxious to dwell upon. There was a time when I very tenaciously held an opinion, at one time promulgated by the late lamented Dr. Buckland, in his 'Reliquiæ Diluvianæ,' to the effect that the bones in caves were relics of the Deluge. That opinion I believe to be quite untenable. Not only did different causes operate in producing similar phenomena, but also there is overwhelming evidence that they were formed at different times. Some, as we have seen, were dens of wild animals; others, places of human abode or sepulture; others, again, mere drains; while some can boast that they entomb animals which have long ceased to exist—'the giants of those days.' This is by no means a general rule.

The fact found to prevail so extensively, and so confidently appealed to, namely, that all bore marks of the action of water, is a mere consequence of the course of their existence;—if water did not frequently run in great quantities where they are found, they never would have been there at all. Revelation is, however, much better without such equivocal support as misinterpreted facts. It can well spare this testimony, since Science has laid nearly all her latest and most glorious laurels at its feet. What we should never have looked for, namely, the marks of an inundation which only lasted a year many thousand years ago, has not been found. But its very absence might be cited as a corroborative fact. Let us, however, at least congratulate ourselves that Geology displays as much the wonders of the Creator as its sister sciences, Chemistry, Mineralogy, or Botany, and they bewilder us with visions of God's immensity. These silent caves, never for ages past enlivened by the busy hum of life, scarcely echoing to the footsteps which explore their hidden beauties, have within themselves a wondrous record of this planet's changes.

Geologists have been accused of requiring too much time for the operation of the mutations they have helped to disclose; but look upon this architecture—this glorious tracery of Nature—remembering that it has been formed atom by atom, and line by line; consider how long it must have taken a mere drop of water to take down from above the marvellous columns which adorn this palace of stone, and ask, Will years, even counted by hundreds, cover the period it includes?

Man, in his busy speculations among the stars, has told of wondrous things. He has pointed out orbs whose distance from us he has discovered, but his numbers have an unmeaning sound, which his own mind cannot reach. He has traced dim clouds to universes whose existence may have finished since the radiance which now shines upon him proceeded from them. All his discoveries enlarge our small ideas of the immensity of Omnipotence. And does not Geology do the same? Beneath the soil, carpeted by various flowers which herald forth the beauty of a world to come, are secrets which are only known to man in part.

But these revelations, small as they are, stretch far beyond his comprehension. He learns that the dust he treads upon was once alive, that the rock on which he takes his stand has lived and died— has been a thing of life, and is now a stone: and this is a time which reaches so far back as only to be understood by Him who was from eternity. He sees that a cavity (but an atom in the world) has, by the small dropping of water, created itself into a palace, and then has it stood a silent witness to the earth's history, has become a cemetery of a creation swept away in one of its changes. But this is not all, nor even a part. It requires now a laborious man to learn all which, little by little, has been revealed to those who have looked into the past history of creation; and man, pausing in his vain endeavour to stretch his mind to the capacity of that which has no bounds, is obliged to rest himself from the thought of the Infinite, and to confess that, whether he searches in earth, or sky, or sea, he is everywhere met by the visions of the Illimitable.

CHAPTER XIII.

CAVES.

CAVES.— MOUNT BURR CAVES. — VANSITTART'S CAVE.— MITCHELL'S CAVE. — THE DROP-DROP. — BONES OF A LARGE KANGAROO. — ELLIS'S CAVE.— UNDERGROUND DRAINAGE. — CAVES AT LIMESTONE RIDGE.— OTHER CAVES.— CONCLUSION.

I COME now to describe the other caves in the district of which I have undertaken to write. As already repeatedly remarked, where the whole district is one formation, and that a loose limestone, these may be expected to be numerous enough, and so, in fact, they are.

The first intended to be described are those of Mount Burr. This hill, as my readers are aware, is an immense upheaval of limestone by trap rock, causing a fault similar to that of Leake's Bluff, with this difference only, that trap rock is visible on the latter and not on the former. It is a hill covered on all sides with the outcroppings of the limestone, and, towards its base, has several little escarpments. Some of them have troughs, or small valleys, descending to their base from the higher land beyond them, and then any drainage which comes down either lies as a pond at the foot of the rock, or drains underneath it.

A A

One of these places where the water drains has given rise to a fine series of caves. No one would suspect, from the outside, that there was so extensive a cavity within. The limestone appears perforated and honeycombed, of from four or five feet above the ground, but there is only one very small aperture, through which a man can barely creep. It was only lately that the caves were discovered by a person determined to see which way the water drained. On creeping through the orifice, a very large chamber is discovered, with the roof not more than sixteen, or, in places, at most twenty, feet from the ground, but very irregular. There are few or no stalactites, but the water drops through in quantities quite large enough to make them in a very short time, and, therefore, we may conclude that the caves have not been very long in existence.

There are three or four wide passages off this chamber, leading to as many subterranean ones. As they traverse underneath, there are several places where the light comes in from above, through apertures in the limestone. These were noticed long before their connection with the caves were known, and were thought to be natural wells. The whole extent of the caves has not been ascertained, but they have been followed for an immense distance, without diminishing in width or in height. They are not, in other respects, very remarkable or beautiful, as they contain but few stalactites, and, as far as they are yet known, no other natural curiosities. It is true that their aspect is both grotesque and singular, having a very wild appearance by the light of the torches necessary to explore them.

Where the roof has support, they have the appearance of groined arches; and the fanciful manner in which the water has worn the faces of the stone is like rough tracery. What has a beautiful appearance, and is, in fact, a singular phenomenon, is a series of most delicate wreaths, which hang down from the ceiling like clusters of long silken hair. These are roots of trees, which grow in the limestone above, and descend through cracks and crevices until they reach the floor of the cavern beneath. Some few are as thick as a man's finger, and these, not only descend from above, but grow into the floor beneath, looking like the branches of the banyan tree or iron pillars, planted to support the roof. The majority, however, are thin and silky, and look almost like gossamer. When one takes in hand what appears to be a thick bunch, it proves as light as a feather, composed of thin shreds, which throw out tubers every now and then, which interlace and form a compact network. There are no cracks visible where they are thickest, and yet, that such thin filaments could penetrate the limestone unless there were apertures, does not appear possible. This is the only cave where I observed anything of the kind, even though the limestone ceiling might be thinner, with trees on the top. All the roots were brown in colour, possessing a thin, light cortical substance, and being white inside. They were wet, and this is probably the principal source of moist nourishment to the tree, the soil above being exceedingly dry. Some of the bunches of fibre were at least ten feet long.

Does it not seem wonderful how far the other ingredients necessary for plant life must have been carried to meet the sole want of water, and how almost like an instinct it seems that a tree should send by chance a rootlet into the cave, and, learning that water could be had, kept on adding and increasing the growth until there was a large surface exposed to take advantage of the favourable position? But perhaps it would be more fair to say, that, as the moisture was favourable to growth, it was where the plant could procure it that growth would be first and best promoted.

Another peculiarity in this cave which has not been met elsewhere in this district is, that it is full of mud, about eight inches deep. This renders the exploration of the cave a matter of great difficulty, besides being disagreeable in the extreme. There are pools of water here and there, but they are, for the most part, surrounded and bottomed with finely-levigated mud, which covers the limestone floor. This moisture seems to have arisen from a swamp, which, it would appear, drains into the cave when the rains are very heavy. I never saw mud in any other cave, and the exception, in this case, is in consequence of the extreme lowness of the aperture, whereas, in all other caves, the opening is at a height where only clear water could reach.

At the mouth of the cave there is a breccia of

bones, which have been brought down by a current of water and deposited at the entrance until cemented into the limestone. It was impossible to detach any of these bones without almost completely destroying them. They appeared large bones, very like those of a kangaroo, though only the ends of them were visible. There were no other bones of any kind in or near the cave, with the exception of a few bones of the vulpine phalanger, or opossum of the colonists, which were strewed upon the mud; but this latter was of too great thickness to enable one to explore with facility the limestone underneath.*

The next cavern worthy of notice is that which here goes by the name of Vansittart's Cave. It is a round opening in the ground close to Mount Gambier, about forty feet across, with a very long sloping precipitous path leading to the bottom, covered over with ferns and rank vegetation. The cave is not, properly speaking, entered until the pit is descended to a depth of some seventy feet; then there is a semicircular opening or arch, which goes slanting under the limestone for forty feet more, where water is reached. At the edge of the water there is scarcely light enough to perceive anything, especially the water, which is so wonderfully clear that its interposition between the observer and the floor is not for a long time per-

* From among these bones I have since obtained specimens of bones of the large animals described in the Appendix. They were mixed with those of existing species, and one bone was evidently the spurious molar of the *Macropus Titan* (Owen), an extinct kangaroo of gigantic dimensions, the skull being larger than that of an ox.

ceptible, so that one runs imminent danger of walking into it without knowing whence the moisture proceeds. Up to the water's edge the width of the cave is about twenty feet, but there it suddenly narrows to a mere low passage, which is seen by torch-light to go a great distance farther. The water prevents its complete exploration. This latter deepens rapidly from the side, which, at the distance of about twelve feet, is five-and-twenty feet deep, and yet, even here, such is the clearness of the water, that every object on the bottom is clearly seen. A gentleman who visited the cave a short time since was very anxious to ascertain what might be the length of the aperture, but, after swimming a short distance, the intense cold compelled him to return, without much more information than he could have gained from the side. I imagine the water to belong to the general water-level of the whole district, as the wells are all about ninety feet deep here. At one time, however, it must have been lower for these passages to be hollowed out, and very likely the cave was occasioned by a drainage from the small hills at some little distance from the mouth of the cave.

There were no bones here at all perceptible. The entrance is surrounded with an abundance of the small fern, *Asplenium laxum*, an acrogen which is not found anywhere in the neighbourhood, though the *Pteris esculenta* and *Adiantum assimile* abound here. There is also a cave at no great distance, and which is so small as to demand no further notice, in which the fern-tree grows. There is no other of the kind (*Cibotium Billardieri*) in the neighbourhood, and yet one of the plants reaches from the foot of the cave to the summit, and seems to reach its mouth.

We pass on now to Mitchell's Cave, close to the one we have just been describing. It is a hole very much like the opening to the preceding, except that the bottom is reached by a winding path, and then opens into a chamber at right angles to the diameter of the entrance. It has a pool of water shelving under the rock, which is so deep as to give it, clear as it is, a deep sea-blue tint. There is no mark of any passage continuous with the cavern, but the roof, where a section of it is seen, is much honey-combed, and must have contained many passages for water. Evidently these were all covered over at one time, and the present cavern was only exposed, within a comparatively recent period, by the falling in of the roof. There are also a few sand-pipes visible in the sections exposed, of a width varying from two feet to a few inches. None of these descend through the strata into the cave, and they are all filled with the red ochreous sand which here results from the decomposition of the limestone. This cave is remarkable as having been one of the sole reservoirs of water for the early settlers before any wells were sunk; now, however, it is little used for the purpose, and is enclosed as a Government reserve. The water is full of a cypris and cyclops, the shells of which seem to strew the bottom. There is also much conferva, a shrimp-like brachiopod, and a minute paludina, which seem to blacken the water, and they cover a piece of wood very soon after its immersion.

At about four miles from this place there is another remarkable cavern, called the Drop-Drop, from the circumstance of water dripping from above into it. It thus formed, at one time, a constant supply of water to those who lived in its vicinity. The place is not remarkable, except for being long and narrow, and going a very great depth under ground. In the neighbourhood there are a very large number of this sort of caves, very richly supplied with stalactite. They are being dug into every day, as wells are being sunk and the ground tilled. Indeed, the resonance of all the hills in the locality shows the ground to be completely undermined. On one occasion, a dray and bullocks fell bodily into a cavity of this description, their great weight having broken through the roof.

It was in one of these cavities that a bone breccia was found, where, under a small aperture, about two feet wide, was a mass of translucent limestone, in which bones were embedded. These must have fallen in from above, as there was no drainage to the mouth of the cave, which was, besides, not ramified, but a chamber about thirty feet deep and fourteen wide. The bones were all of a species of kangaroo existing in the neighbourhood, and were embedded together in rather an indiscriminate manner. The specimens I saw were mostly jaw-bones.

On removing this breccia, one of much older date was found beneath. From this I procured one remarkable bone, probably belonging to a species of kangaroo called the Euro, which is only found 400 miles to the north of Adelaide, or 700 miles from where it was found. From the engraving it is seen that, while the animal must have had a much more massive frame than the ex-

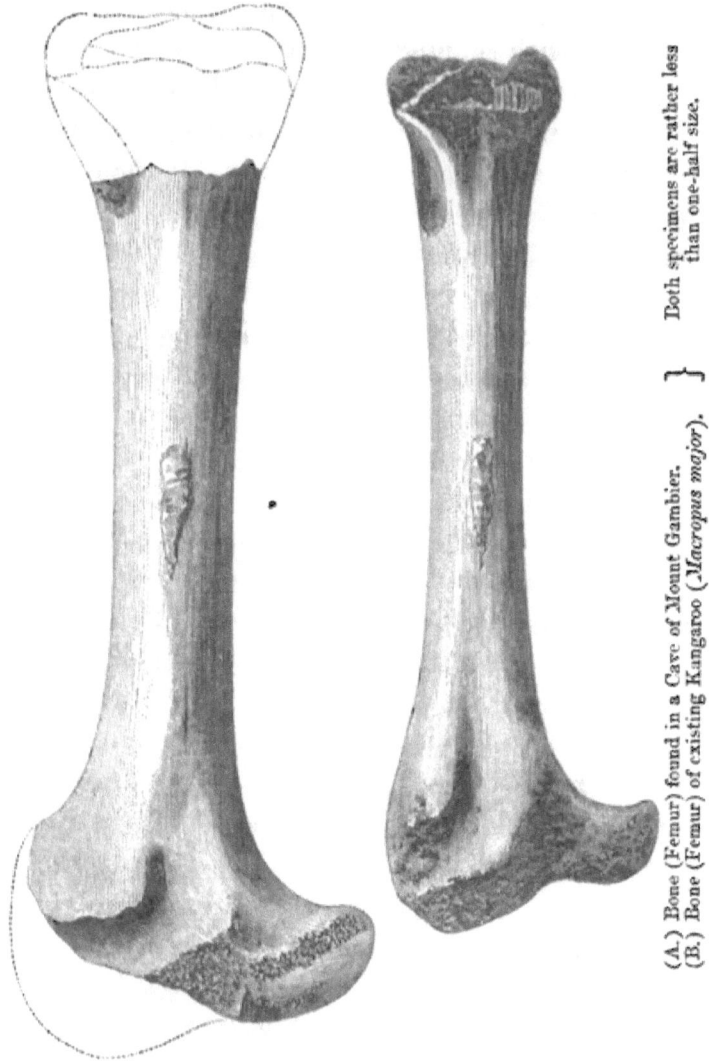

(A.) Bone (Femur) found in a Cave of Mount Gambier.
(B.) Bone (Femur) of existing Kangaroo (*Macropus major*).
} Both specimens are rather less than one-half size.

isting kangaroo of the neighbourhood, they were shorter and heavier, and much less fitted for speed. In fact, before I knew the qualities of the Euro, I concluded that this bone was the femur of a kangaroo, evidently a much larger animal than that which we have around us at present, but not possessed of such running or jumping powers. The length of the bone showed that the leverage could not be great; and it was after this that I heard of the Euro possessing these characters. In the bone, the depth of the introchanteric fossa is very remarkable. The animal matter was entirely absent, and the specimen extremely light for its size.

The occurrence of this bone inclines us to speculate on the causes of the banishment of the animal from this quarter, and its being only found at a much warmer locality. It is rarely to be met with anywhere, and this may be because, from its low powers of running, it became a more easy prey to the aborigines, or wild dogs, while its large size made it a much more desirable prey. This, perhaps, is the only instance where the bones found in caves, apparently larger than of any existing species, have been found to have representatives still existing. Probably the bones found at Wellington Valley, spoken of by Sir Charles Lyell in his 'Manual of Geology,' may have been those of the Euro.* But that only one bone of an animal is found which possibly was very common in former times in the same locality, shows how very few of terrestrial faunæ get embedded in strata and leave

* See Appendix.

records of their existence, and, therefore, how very weak is negative evidence with reference to the former state of the earth's surface. We can no more infer the character of animal life from the absence of certain remains, than we could guess all the animals of an island from a few species brought home by a naturalist who made a small collection during a short visit.

Now comes the account of another cave, which differs most materially from all that have been previously described. Close to Mr. Ellis's station, within about five miles of Mount Gambier, there is a whim erected over a small hole in the rocks. Underneath this, at the depth of about seventy feet, there is a long passage or cavern, through which a deep stream of water flows. It has been followed in a boat, without the passage becoming more narrow or the water more shallow, and very likely continues till near the coast, where, as before mentioned, there are several natural springs, where large quantities of water boil through the limestone rock. In spring and summer, there is a distinct stream or ripple visible on the surface, as seen from the top of the well. Doubtless this is one of the many passages through which the surface-water drains from this district. It had long been a subject of speculation how the water drained from the southern part of this country. About that of the northern part there was but little difficulty. The creek which drains the flat near the caves, as well as some other drainage, goes into a large swamp, on the Mosquito Plains, known as the Mosquito Swamp. This, when full, drains into the Salt Creek, in a north-westerly direction, and this creek into the Coorong, and thence into the Murray. But accounting for the southern drainage is not so easy, and the only way of explaining for the disappearance of the excess of water is to suppose that it drained under ground. This was corroborated by several facts. In many of the wells in this district a distinct ripple is at one time observable on their surface, and floating objects placed on one side are rapidly borne to the other. I have even heard persons say, who resided where the depth of the water-level is not great and the limestone rock cropping out, that they could distinctly hear a sound underneath them like the rolling of water. There is a swamp, near Mount Graham, at the head of the Reedy Creek, whose sides are surrounded, here and there, with out-cropping limestone rock. When the swamp overflows, the water drains under these rocks, which are much honeycombed at these places, though there is no appearance of caves. It can be heard rumbling away at a distance.

This cave, then, at Mr. Ellis's, is probably one of the channels of drainage. Doubtless it is supplied by many small streams which merge to this point, and its continued action has hollowed out the passage where it runs. Its course is about southeast, and either it comes to the surface in one of the numerous fresh-water springs which abound on the coast, or else it comes up under the sea, like the water resulting from the katavothra, in Greece.

Very likely, in the course of time, other passages like this will be found, and some of those empty galleries, which are now so frequently dug into at a small depth from the surface, are beds of streams, which the upheaval of the land has deprived of their office. Should this upheaval continue, the passage we are now treating of will eventually become dry.

There is another very remarkable cave, about three miles from Mount Shanck. It is full of water, and soundings made from the side with sixty feet of line found no bottom. It would be interesting to know the nature of the bottom, as probably an approximate guess of the thickness of the limestone strata might then be arrived at, for there is some considerable distance from the entrance of the cave to the water's edge.

Next in interest to this is a series of caves at the Limestone Ridge Station, a little over the boundary near the Victorian township of Munbannar.* This locality is full of caves, most of them leading into one another by tortuous passages made by beautiful stalactites. The description of one cave, however, is so very like another, that I fear I should grow tedious were I to enlarge much upon their varieties. It will be sufficient, therefore, to say, that the locality possesses about twenty, within

---

* This is the native name of the place, and, like most native names, is rather euphonious. It is a pity that so few have been preserved. As a sample of their musical sound, we might cite a few which would be infinitely preferable to transplanted British names, such as Liverpool, Newcastle, &c., which will, in time, produce endless geographical confusion. Caramedulla, Aldinga, Yankallilla, Lillimer, Kaniver, Pareene, &c., are infinitely preferable, and such names as these are not the best specimens.

a short distance of each other.* There are none deeper than about twenty feet, none very wide, and they all seem to be connected with each other by winding passages. It is one of these which contains a chamber only open to the sky by a small aperture, and this nearly perfectly concealed. Underneath is a heap of bones—a melancholy monument to those unfortunate kangaroos who, prior to leaping, did not take the precaution of looking. The chamber around is also covered with bones, as mentioned in a previous chapter.

There are also many other caves, which are hardly worth a minute description, now that the leading features of the most important have been described. There is, for instance, a cave at Mr. Meredith's station which is a mere vault, with tumbled boulders on the floor, and many narrow passages, nicely decorated with stalactite; one at Mr. Johnstone's station (Mount Muirhead), which is a very plain cavity, with two entrances; one at Mr. Ellis's, which is entered by a very rapid descent, leading into a lofty vault. There are, besides, a great many more, but far too numerous to particularise here. The three last are on higher ground, and therefore connected with sudden flows of water. None of them are near creeks, but probably may have been hollowed out by the floods which followed the eruptions in the southern part of the district.

* One is such a famous resort for bats, that it is called the Bat Cave in consequence. They frequently extinguish the lights of explorers, and in their screechings and fluttering remind one of Humboldt's Guacharos.

CHAPTER XIV.

CONCLUDING REMARKS.

MY observations on this district are, for the present, brought to a close. They form a sketch, and a very imperfect one, of what has been observed in this part of the world, which, however remote, does not seem to have come to its present state of things by a very different process from what has happened elsewhere. Much will remain to be done by future observers, either by making new observations and collecting new facts, or by extending the application of those already observed. There may be some slight utility in what has been sketched in the preceding pages. Like the 'Natural History of Selborne,' it has been the occupation of many a passing hour in the Bush, where amusements are otherwise few; and, though it may appear to go unnecessarily into detail, it may, like the same work, be made the groundwork of larger and more general conclusions. Everyone has it in his power to contribute, in some degree, to the world's stock of knowledge. If this were acted upon, the different sciences would soon assume other aspects; and I cannot think that the small details which are food to a speculative mind are ever dry to those who seek for information.

The conclusions to be drawn from what has been stated in the preceding pages, though reducible to small compass for actual results, are not uninteresting, and may lead to something more important. They may be described as follows:—

I. There has been in Australia an immense area of subsidence during the Pleiocene period, at a time when Rome, parts of Italy, Vienna, and parts of Austria, Piedmont, and Asia Minor were under the sea.

II. This subsidence was accompanied by a coral formation, very similar to the subsiding area of the Pacific at the present time, and, though all the appearances are those of a reef of true zoophytic corals, the predominant fossil is a massive *Cellepora*, while true corals are rare.

III. This gives rise to the suspicion that *Bryozoa* may build reefs and atolls as well as true corals.

IV. That the subsidence ceased, and probably about that time volcanic disturbance commenced, and gave rise to submarine craters.

V. That, after the cooling of the lava from these submarine craters, a deposit of small fragments of shells was thrown down from an ocean current.

VI. That this became hardened into stone, and was then upheaved from the sea, during which process large portions of it became washed away.

VII. That the latter part of the upheaval was separated by a long lapse of time from the subsidence, because the latter strata show some difference in their fauna.

VIII. That while upheaval was going on, until very recently, extensive volcanic disturbance took place, giving rise to craters which are all now extinct.

IX. That the upheaval coralline rock, when decomposed, has given rise to a very indifferent sort of soil, of a sandy character, which causes large tracts of arid useless country in this part of Australia.

X. That the same rock, being of a loose texture, easily allowed water to percolate through, forming caves and underground passages, besides honeycombing the ground in all directions.

XI. That, while these operations proceeded, the animal life was of a slightly different character from what is found in the same locality now, though, probably, the land animals were not specifically different from individuals in other parts of the Australian continent.

These numerous changes seem to have taken place without any very vast convulsion of nature, or phenomena different from what happen in the world now. It has been the custom, lately, to say this of all the operations of Geology. No one, however, who has studied the question, will deny that there are peculiar characters in different geological epochs which indicate something very diverse from the character of the earth's surface now. Thus, there is the age of the Silurian slates, enormous masses of finely-levigated mud, derived from whence we know not, and very sparingly supplied with animal remains: the carboniferous era, with the enormous swamps of fern vegetation; the Wealden, with its gigantic reptiles; the chalk, with its corals and corallines. And so we may say of our crag deposits. They give evidence of a peculiar state of things, and seem in every case to have been followed by the same results. Geology is like history—its events repeat themselves, but not the same events, and each period has a character which seems to have affected the whole earth for the time being.

But even these conclusions must be modified by remembering how many of them rest on negative evidence. The very circumstances under which certain deposits are found may, in securing their own preservation, exclude any but a certain class of organic remains. Thus the coal deposits do not warrant us in concluding that there were no other plants, but rather, where these grew in such abundance, the growth was owing to circumstances which excluded others. Or again, the reptiles which are found in the Wealden mud (the estuary of a former river) may have sought food in such a place, and thus be nearly the only animal embedded. But with every limitation, however, the general character of the fauna or flora, of any period, is always very clearly marked.

All these things show that Geology has some conclusions as certainly established as to enable her to avoid the errors of hasty generalisation. As a science which requires so much from other branches of knowledge, it can ill afford to lose itself in mazy speculations while so much remains to be done. Little by little the edifice is building, and probably small contributions, such as these pages, may be offered without presumption.

It may seem strange that so much food for speculation is to be found in the earth beneath our feet. It leads to much knowledge. Let us not be presumptuous, however. How small it is in comparison with the vast amount still unknown, and yet within reach!—how small by the side of the vast sea of the unknown materials for human knowledge, and how immeasurably insignificant compared with that illimitable knowledge which all eternity will not enable us to understand! Well may I conclude with the beautiful words of an eminent philosopher:—

'Si quid profecerimus, non alia sane ratio nobis viam aperuit quam vera et legitima spiritus humani humiliatio. Quamobrem . . . ad Deum Patrem, Deum Filium, Deum Spiritum, preces fundimus humillimas et ardentissimas . . . ne humana divinis officiant; . . . sed potius ut ab intellectu puro a phantasiis et vanitate repurgato et divinis oraculis nihilominus subdito et prorsus dedititio, fidei dentur quæ fidei sunt; postremo, ut scientiæ veneno a serpente infuso quo animus humanus tumet et inflatur deposito, nec altum sapiamus nec ultra sobrium, sed veritatem in charitate colamus.' *

* Bacon, *Instauratio Magna*.

APPENDIXES.

Appendix No. I.

CAVES AT WELLINGTON VALLEY, NEW SOUTH WALES.

THE following description of the above caves, by Sir Thomas Mitchell, is added to illustrate what has been said in this work on the subject of caves in general:—

'We first descended the fissure at the mouth of the large cave, and then clambered over large rocks until, at 125 feet from the entrance, we found these inequalities to be covered by a deep bed of dry reddish dust, forming an even floor. This red earth lay also in heaps under lateral crevices, through which it seemed to have been washed down from above. On digging to a considerable depth at this point, we found a few fragments of bone, apparently of the kangaroo. At 180 feet from the mouth is the largest part of the cavern, the breadth being twenty-five feet, and the height about fifty feet. The floor consisted of the same reddish earth, but a thick stalagmitic crust extended, for a short distance, from a gigantic stalactite at the farther end of the cavern. On again digging several feet deep into the red earth here, we met with no lower layer of stalagmite, nor any animal remains.

'On a corner of the floor, behind the stalactite, and nearly under a vertical fissure, we found a heap of dry white dust, into which one of the party sank to the waist.

'Passing through an opening to the left of the stalactite, we came upon an abrupt descent into a lower cavern. Having reached the latter, with some difficulty, we found that its floor was about twenty feet below that of the cavern above. It was equally level, and covered to a great but unascertained depth with the same dry red earth, which had been worn down about five feet, in a hollow or rut.

'A considerable portion of the farthest part of the floor was occupied with white dust or ashes, similar to that found in the corner of the upper floor.

'This lower cavern terminated in a nearly vertical fissure, which not only ascended towards the external surface, but descended to an unascertained depth beneath the floor. At about thirty feet below the lowest part of the cavern, it was found to contain water, the surface of which I ascertained was nearly on a level with that of the river Bell. Having descended by a rope, I found that the water was very transparent, but unfit to drink, having a disagreeable brackish flavour.

'This lower cavern is much contracted by stalactites and stalagmites.

'After having broken through some hollow-sounding portions, we entered two small lateral caverns, and in one of these, after cutting through about eight inches of stalagmitic floor, we discovered the same reddish earth. We dug into this deposit also, but discovered no pebbles or organic fragments, but, at the depth of two-and-a-half feet, met with another stalagmitic layer, which was not penetrated. This fine red earth or dust seemed to be a sediment that was deposited from water which stood in the caves, about forty feet below the exterior surface; for the earth is found exactly at that height, both towards the entrance of the first cavern and in the lateral caverns.

'That this cave had been enlarged by the partial sinking of the floor is not improbable, as broken stalagmitic columns and pillars, like broken shafts, and once probably in contact with the roof, are still apparent.

'Eighty feet to the westward of this cave is the mouth of another of a different description. Here the surface consists of a breccia, full of fragments of bones; and a similar compound, confusedly mixed with large blocks of limestone, forms the sides of the cavity. This cave presents, in all its features, a striking contrast to that already described.

'Its entrance is a sort of pit, having a wide orifice, nearly vertical, and its recesses are accessible only by means of ladders and ropes.

'Instead of walls and a roof of solid limestone rock, we found shattered masses, apparently held together by breccia, also of a reddish colour, and full of fragments of bones. The opening in the surface appears to have been formed by the subsidence of these rocks, at the time when they were hurled down, mixed with breccia, into the position which they still retain. Bones were but slightly attached to the surface of this cement, as if it had never been in a very soft state, and this we have reason to infer, also, from its being the only substance supporting several large rocks, and, at the same time, keeping them asunder. On the other hand, we find portions of even very small bones, and also small fragments of the limestone, dispersed through this cementing substance or breccia.

'The pit had been first entered, only a short time before I examined it, by Mr. Rankin, to whose assistance in these researches I am much indebted. He went down by means of a rope to one landing-place, and then, fixing the rope to what seemed a projecting portion of rock, he let himself down to another stage, where he discovered, on the fragment giving way, that the rope had been fastened to a very large bone, and thus these fossils were discovered. The large bone projected from the upper part of the breccia, the only substance which supported as well as separated several large blocks, and it was covered with a large tufaceous incrustation resembling mortar. No other bone of so great dimensions has since been discovered within the breccia.

'From the second landing-place we descended through a narrow passage, between the solid rock on one side and huge fragments, chiefly supported by breccia, on the other, the roof being also formed of the latter, and the floor of loose earth and stones. We then reached a small cavern, ending in several fissures, choked up with the breccia. One of these crevices terminated in an oven-shaped opening in the solid rock, and was completely filled, in the lower part, with soft red earth, which formed also the floor in front of it, and resembled that in the large cavern, already described.

'Osseous breccia filled the upper part of this small recess, and portions of it adhered to the sides and roof adjoining, as if this substance had formerly filled the whole cavity. At about three feet from the floor, the breccia in this cavity was separated from the loose earth below by three layers of stalagmitic concretion, each about two inches thick and three apart; they appeared to be only the remains of layers, once of greater extension, as fragments of stalagmite adhered to the sides of the cavity. The spaces between what remained of these layers were most thickly encrusted with tufaceous matter; those in the upper surfaces, on the contrary, were very white, and free from the red ferruginous ochre which filled the cavities of those in the breccia, although they contained minute transparent crystals of carbonate of lime.

'On digging into the soft red earth, forming the floor of this recess, some fragments of bone, apparently heavier than those of the breccia, were found, and one portion seemed to have been gnawed by a small animal.

'We obtained also in this earth the last phalange of the greatest toe of a kangaroo, and a small water-worn pebble of quartz. By creeping about fifteen feet under a solid mass of solid rock—which left an opening less than a foot-and-a-half above the floor, we reached a recess about fifteen feet high and twelve feet wide. The floor consisted of dry red earth, and, on digging some feet down, we found fragments of bones, a very large kangaroo tooth, a large tooth of an unknown animal, and one resembling some fragments of teeth found in the breccia.

'We next examined a third cave, about 100 yards to

the westward of the last described. The entrance, like that of the first, was tolerably easy, but the descent over the limestone rocks was steeper, and very moist and slimy; our progress downwards was terminated by water, which probably communicated with the river Bell, as its level was much lower when the cave was first visited, during a dry season. I found very pure iron ochre in some of the fissures of this cavern, but not a fragment of bone.

'Perceiving that the breccia where it occurred extended to the surface, I directed a pit to be dug on the exterior, about twenty feet from the mouth of the cave, and at a part where no rocks projected. We found that the hill there consisted of breccia only, and was harder and more compact than that in the cave, and abounded likewise in organic remains.

'Finally, I found on the summit of the same hill some weathered blocks of breccia, from which bones protruded, and a large and remarkable specimen.

'Other caverns containing breccia of the same description occur in various parts within a circuit of fifty miles, and they may probably be found throughout the limestone country not yet examined.

'On the north bank of the M'Quarrie, eight miles east from the Wellington Caves, and at Buree, about fifty miles to the southward of them, I found this breccia at considerable depths, having been guided to it by certain peculiar appearances of subsidence and disruption, and by yawning holes in the surface, which previous experience had taught me to consider as indications of its existence.

'On entering one of these fissures, from the bed of the little stream near Buree, and following to a considerable distance the subterraneous channel of a rivulet, we found a red breccia, containing bones as abundantly as that of Wellington Valley. It occurred, also, amidst masses of broken rocks, between which we climbed until we saw daylight above; and, being finally drawn out with ropes, we emerged, near the top of a hill, from a hole very similar in appearance to the mouth of the cave at Wellington, which it also resembled in having breccia, both in the sides of the orifice and in the surface around it.

'At Molong, thirty-six miles east of Wellington Valley, I found some concreted matter within a small cavity of limestone rock on the surface, and, when broken, it proved also to be breccia containing fragments of bone.

'It was very difficult to obtain any perfect specimens of the remains contained in the breccia; the smallest of the various portions brought to England have, nevertheless, been carefully examined by Professor Owen, at the Hunterian Museum, and I have received from that distinguished anatomist the accompanying letter, containing the results of those researches and highly important determinations, by which he has established several points of the greatest interest, as connected with the natural history of the Australian continent :—

"Royal College of Surgeons,
May 8th, 1838.

"Dear Sir,

"I have examined, according to your request, the fossil remains which you discovered in Wellington Valley, Australia, and which are now deposited in the Museum of the Geological Society; they belong to the following genera:—

"MACROPUS *Shaw.*

"Sp. 1. *Macropus Athos* (Owen).—This must have been at least one-third larger than *Macropus major*, the largest known existing species: it is chiefly remarkable for the great size of its permanent spurious molar, in which respect it approaches the subdivision of Sparo's genus, called *Hypsiprymnus* by Illiger. The remains of this species consist of a fragment of the right ramus of the lower jaw.

"Sp. 2. *Macropus Titan* (O.).—I give this name to an extinct species as large as the preceding, but differing chiefly in the smaller size of the permanent spurious molar, which, in this respect, more nearly corresponds with the existing *Macr. major*. The remains of this species consist of a fragment of the right ramus of the lower jaw.

"In both the above specimens the permanent false molar tooth is concealed in its alveolus, and was discovered by removing part of the substance of the jaw, indicating the nonage of the individuals.

"A portion of cranium with the molar series of teeth of both sides. This specimen I believe to belong to *Macropus Titan*.

"The permanent false molar, which is also concealed in this upper jaw, is larger than that of the lower jaw of *Macr. Titan*, but I have observed a similar discrepancy in size in the same teeth of an existing species of *Macropus*.

"To one or other of the two preceding gigantic species of kangaroo must be referred —

"II. (*a*) Crown of right inferior incisor.

"II. (*b*) Lower extremity of right femur.

"II. (*c*) Lower extremity of right femur, with the epephysis separated, showing its correspondence in age with the animals to which the fossil jaws belonged.

"II. (*d*) Fifth lumbar vertebra.

"II. (*e*) Tenth or eleventh caudal vertebra. The proportion of this bone indicates that these kangaroos had a relatively stouter and perhaps shorter tail than the existing species.

"*Macropus sp. indeterm.*—Agrees in size with *Macropus major*, but there is a difference in the form of the sacrum, the second vertebra of which is more compressed. To this species, which cannot be determined till the teeth be found, I refer the specimens marked

"III Sacrum. III. (*a*) Proximal end of left femur. III. (*b*) Proximal end of left tibia, in which the anterior spine sinks more gradually into the shaft than in *Macr. major*. As this is the only species with the skeleton of which I have been enabled to compare the preceding fragments, I am not able to pronounce as to their specific distinctness from other existing species of equal size with the *Macropus major*.

" *Macropus sp. indeterm.*—From want of skeletons of existing species of kangaroo, I must leave doubtful the specific determination of a species smaller than *Macropus major*, represented by the left ramus of the lower jaw, in which the permanent false molar is in place together with four true molars, and which would therefore be a species of *Halmaturus* of Fred. Cuvier.

" *Macropus* (5).—Part of the left ramus of the lower jaw, with two grinders in place, and a third which has not quite cut through the jaw.

" V. (*a*) Sixth and seventh grinders, according to the order of their developement, right side, upper jaw of a kangaroo not quite so large as *Macropus major*.

" Several other bones and portions of bone are referable to the genus *Macropus*, but they do not afford information of sufficient interest or importance to be specially noticed.

" Genus HYPSIPRYMNUS.

" *Hypsiprymnus sp. indeterm.*— A portion of upper jaw and palate, with the deciduous false molar and four true molars in place on each side; the fifth or posterior molar is concealed in the alveolus, as also the crown of the permanent false molar.

" *Hypsiprymnus.*—Part of the right ramus of the lower jaw, exhibiting a corresponding stage of dentition:

" *Obs.* This species is rather larger than any of the three species with the crania of which I have had the opportunity of comparing them: there is no evidence that it agrees with any existing species.

" Genus PHALANGESTA.

" No. 7. Cranium coated with stalactite.

" No. 7 (*a*). Part of right ramus, with spurious and second molar.

" No. 7 (*b*). Right ramus, lower jaw.

" *Obs.* The two latter specimens disagree with the *Phal. vulpina*, in having the spurious molar of relatively smaller size, and the second molar narrower; the symphysis of the lower jaw is also one line deeper in the fossil. As the two latter specimens agree in size with the cranium, they probably are all parts of the same species, of which there is no proof that it corresponds with any existing species.

"But a comparison of the fossils with the bones of these species (which are much wanted in our osteological collection) is obviously necessary to establish the important fact of the specific difference or otherwise of the extinct phalanger.

"Genus PHASCOLOMYS.

"Sp. *Phasc. Mitchellii.*—Mutilated cranium.

"No. 8 (*a*). Part of lower jaw belonging to the above.

"No. 8 (*b*). Right series of molar teeth *in situ*.

"No. 8 (*c*). Right ramus of the lower jaw.

"*Obs.* These remains come nearer to the existing species than do those of any of the preceding genera; but, after a minute comparison, I find that there is a slight difference in the form of the grinders, which, in the fossil, have the antero-posterior diameter greater in proportion than the transverse; the first grinder is also relatively larger and of a more prismatic form; the upper incisors are less compressed and more prismatic. This difference is so well marked, that, once appreciated, any one might recognise the fossil by an incisor alone. There is a similar difference in the shape of the lower incisor. The fossil is also a little larger than the largest wombat cranium in the Hunterian collection. From these differences I feel no hesitation in considering the species to which these fossils belong as distinct, and propose to call it *Phascolomys Mitchellii.*

"Genus DIPROTODON.

"I apply this name to the genus of Mammalia represented by the anterior extremity of the right ramus of the lower jaw, with a single large procumbent incisor. This is the specimen conjectured to have belonged to the dugong, but the incisor resembles the corresponding tooth of the wombat in its enamelled structure and position; but it differs in the quadrilateral figure of its transverse section, in which it corresponds with the inferior incisors of the hippopotamus.

"Genus DASYURUS.

"*Das. lancarius* (O).—I apply this name to the species to which the following remains belong:—

" XI. Portions of the left side of the upper jaw.

" XI. (*a*) Ditto.

" XI. (*b*) Left ramus of lower jaw, with lost grinders.

" XI. (*c*) Anterior part of the right ramus of lower jaw.

" This species closely resembles *Das. ursinus*, but differs in being one-third larger, and in having the canines or laniaries of proportionately larger size.

" The position of the teeth in the specimen marked XI., which are wider apart, leads me to doubt whether it is the lower jaw of *Das. lancarius*, or of some extinct marsupial carnivora of an allied but distinct species.

" The general results of the above examination are:—

" 1st. That the fossils are not referable to any known extra-Australian genus of Mammals.

" 2nd. That the fossils are not referable, from the present evidence, to any existing species of Australian Mammals.

" 3rd. That the greater number certainly belong to species either extinct or not yet discovered living in Australia.

" 4th. That the extinct species of *Macropus, Dasyurus, Phascolomys*, especially *Macr. Athos* and *Macr. Titan*, are larger than the largest known existing species.

" 5th. That the remains of the saltatory animals, as the *Macropi, Helmaturi,* and *Hypsiprymni*, are all of young individuals; while those of the burrowing wombat, the climbing phalanger, and the ambulatory dasyure are of adults.

" I remain, dear Sir, &c.,
"(Signed)   RICHARD OWEN."

'To this it may be added, that the wombat's skull is fully as large as the skull of an elephant.

'Nothing could be discovered, in the present state of these caverns, at all likely to throw any light on the history or age of the breccia, but the phenomena they present seem to indicate more than one change in the physical outline of the adjacent regions, and probably of more distant portions of Australia, at a period antecedent to the existing state of the country.

'Dry earth occurred in the floor of both the caverns at Wellington Valley and in the small chamber of the breccia cave; it was found, as before stated, beneath the three lines of stalagmite and the osseous breccia. It seems probable, therefore, that this earth once filled the cave also to the same line, and that the stalagmite then extended over the floor of red earth. Moreover, I am of opinion that the interval between the stalagmite and the roof was partly occupied by the bone breccia, of which portions remain attached to the roof and sides above the line of stalagmite. It is difficult to conceive how the mass of red earth and stalagmitic floors could be displaced, except by a subsidence in the original floor of the cave. But the present floor contains no vestiges of breccia fallen from the roof, nor any remains of the stalagmitic crust once adhering to the sides—which are both, therefore, probably deposited below the present floor. In the external or upper part of the same cave, the floor consisted of the red dust, and was covered with loose fragments of rock, apparently fallen from conglomerated masses of limestone and breccia, which also, however, extended under the red earth there. Thus it would appear that traces remain in these caverns, first, of an aqueous deposit in the red earth found below the stalagmite in one cavern, and beneath breccia in the other; secondly, of a long dry period, as appears in the thick crust of stalagmite, covering the lowest deposit in the largest cavern, and during which some cavities were filled with breccia, even with the external surface; thirdly, of a subsidence in the breccia and associated rocks; and lastly, of a deposit of red earth similar to the first.

'The present floor in both caves bears all the evidence of a deposition from water, which probably filled the interior of the cavern to an unknown height. It is clear that sediment deposited in this manner would, when the waters were drawn off, be left in the state of fine mud, and would become, on drying, a more or less friable earth. Any water charged with carbonate of lime, which might have been subsequently introduced, would have deposited the calcareous matter in stalactites or stalagmites; but the general absence of these is accounted for in the dryness of the caves. This sedimentary floor contained few or no bones, except such as had previously belonged to the breccia, as was evident from the minuter cavities having been still filled with that substance.

'I do not pretend to account for the phenomena presented by the caverns, yet it is evident, from the sediments of mud forming the extensive margins of the Darling, that at one period the waters of that spacious basin were of much greater volume than at present; and it is more than probable that the caves of Wellington Valley were twice immersed under temporary inundations. I may, therefore, be permitted to suggest, from the evidence I am about to detail of changes of level on the coast, that the plains of the interior were formerly arms of the sea, and that inundations of greater height have twice penetrated into or filled with water the subterraneous cavities, and probably, on their recession from higher parts of the land, parts of the surface have been altered and some additional channels of fluviatile drainage hollowed out. The accumulation of animal remains, very much broken and filling up hollow parts of the surface, show, at least, that this surface has modified since it was first inhabited, and these operations appear to have taken place subsequently to the extinction, in that part of Australia, of the species whose remains are found in the breccia, and previously to the existence in at least the same districts of the present species.

'No entire skeleton has been discovered, and very rarely were any two bones of the same animal found together. On the contrary, even the corresponding fragments of a bone were frequently detected some yards apart. On the other hand, it would appear, from the position of the teeth in one skull, that they were only falling out from putrefaction at the time the skull was finally deposited in the breccia, and from the nearly natural position of the smaller bones in the foot of a *Dasyurus*. It can scarcely be doubted that this part of the skeleton was embedded in the cement when the ligaments still bound the bones together. The united radius and ulna of a kangaroo are additional evidences of the same kind; and yet, if the bones have been so separated and dispersed, and broken into minute fragments, as they now appear in this breccia, while they were still bound together by ligaments, it is difficult to imagine how that could take place under any natural process with which we are acquainted. It may, however, be observed, that the breccia is never found below ground without unequivocal proofs in the rocks accompanying it of disruption and subsidence, and that the best specimens of single bones have been found wedged between huge rocks where the breccia is found like mortar between them, in situations eight or ten fathoms under ground.'

Appendix No. II.

FOSSIL CLIFFS OF THE GREAT AUSTRALIAN BIGHT.

THE following description of the nature of the fossil cliffs of the Australian Bight is taken from the narrative of Mr. Eyre, who, in 1840, made a terrible and disastrous journey round them:—

'Being now at a part of the cliffs where they receded from the sea, and where they had at last become accessible, I devoted some time to an examination of their geological character. The part that I selected was high, steep, and bluff towards the sea, which washed its base, presenting the appearance described by Captain Flinders, as noted before. By crawling and scrambling among the crags, I managed, at some risk, to get at these singular cliffs. The brown or upper portion consisted of an exceedingly hard, coarse, grey limestone, among which some few shells were embedded, but which, from the hard nature of the rock, I could not break out; the lower or white part consisted of a gritty chalk, full of broken shells and marine productions, and having a somewhat saline taste: parts of it exactly resembled the formation that I had found up to the north, among the fragments of table land. The chalk was soft and friable at the surface, and easily cut out with a tomahawk; it was traversed horizontally by strata of flint, ranging in depth from six to eighteen inches, and having varying thicknesses of chalk between the several strata. The chalk had worn away from beneath the hard rock above, leaving the latter most frightfully overhanging, and threatening instant annihilation to the intruder. Huge misshapen masses were lying with their rugged pinnacles above the water in every direction at the foot of the cliffs, plainly indicating the frequency of a falling crag; and I felt quite a relief when my examination was completed, and I got away from so dangerous a post.' *

From this extract, it appears beyond much doubt that the cliffs are the same formation as those of the Murray and those of Mount Gambier. The upper and lower deposits are identical with those of the latter places, and strongly resemble the mode in which the Pleiocene Crag occurs at home. Thus a geological period, which has left but slender records in Europe, is largely represented in Australia, and forms a very large portion of its continent.

* *Expedition to Central Australia.* By E. J. Eyre. London: Boone, 1845.

INDEX.

INDEX.

ACACIA MOLLISSIMA, 30
Acalephæ, or jelly-fishes, 137
Adelaide, position of the city of, 18. Character of the rocks near, 59. Range of hills on which it is built, 110. Strata on which it is built, 208. Earthquakes near, 213
Adelaide, colony of. *See* Australia, South
Adelsberg caves, 321
Adiantum assimile, 358
Albert, Lake, 204, 205
Alexandrina, Lake, 204
Alps, the Australian, 19. Their height, 19, *note*
America, now in her Pleiocene period, 139
Amygdaloidal trap at Grant Bay, 158
Araucarias found only in Norfolk Island and Australia, 139
Arnhem's Land, 15
— geological formation of, 17
Arthrozamiæ, 139
Asia Minor, tertiary strata of, 88
Asplenium laxum at the mouth of Vansittart's Cave, 358
Asteroidea, fossil, 78
Astræa found on the South Australian coast, 187, 188
Astro-Pecten, 83
Atolls, or ring islands, 13, 14. White mud of the, 92. Description of an atoll, 96. Darwin's theory of the formation of, 125. Probable remains of atolls at Swede's Flat and Half-way Gulley, 131. Causes why one side of an atoll is invariably broken down, 149
Australia, geography of, 12. Nature of a new country told by the scenery of the coast, 13. Geological queries to be answered by Australia, 14. Formation of the coast, 15. The continent formerly separated into two halves, 17. Australian Cordillera, 18. The South Australian chain of mountains, 19—21. Traces of the action of glaciers in them, 20. Metals and minerals found in them, 21, 22. Geology of Northern and Western Australia, 22. General view of Australian geology, 24. Meteorology of Australia, 24. Geological connection between Australia and the older hemisphere, 88. Former higher temperature of Australia, 99—134. Upheaval of a portion of Australia now taking place, 135. Australia geologically far behind the rest of the world, 139. Badly adapted for the habitation of man, 141 — 144. Periodically dry seasons, 145. The Australian aborigines extremely degraded and helpless, 146. Aspect of the coast of Australia, 182. Upheaval of the coast of Australia, 207. Absence of active, and few extinct, volcanoes in Australia, 225. Specimen of the beauty of the Australian flora, 267. Beasts of prey, 313.

Australia, South, ignorance respecting its geology, 8. Mr. Selwyn's catalogue of its rocks, 9. Its formation and mineralogical productions, 9. The South Australian range of mountains, 19—21. Description of the South-

Eastern District, on the surface, 26. And of the rocks, 58. Table of fossils found in, 77. Extent of the South-Eastern District, 103. Boundary line between South Australia and Victoria, 120. Perils of the coast from reefs, from Rivoli Bay to Guichen Bay, 162. Sand-drifts round the coast, 166. Antagonistic forces at work on the coast, 170. Destruction of the cliffs in winter, 170. What has become of the detritus? 171. History of the deposit as presented by the rocks, 171. Origin of the sand of the Australian coast, 187. Lakes on the coast, 195. Upheaval of the coast, 205. Proved from the coast-line, 206. And from the rivers, 208. Earthquakes, 213. Periods of rest, 219. South Australian volcanoes, 224. Rich meadow-like appearance of the country between Mounts Gambier and Shanck, 264. The smaller volcanoes, 282. Connection between them, 288. Gold in South Australia, 298. Granite in the bed of the Murray river, 298. Caves in South Australia, 299, et seq.

Australia, Western, coal beds of, 22. Little known of the geology of, 22

BAHIA BLANCA, sand dunes of, and sandstone near, 222
Bald Head, coral found on, 115, 116. Supposed fossil trees at, 165
Ballarat, rise of, 4
Bandicoot, bones of, found in the first cave at Mosquito Plains, 335—338
Banksia integræfolia, of the ridges, 30. Of the Honeysuckle Country, 42
Banksia ornata, 36
Bark, stringy, 31
Barossa Mountains, 110
Barrier reef, the great, of the west side of Australia, 22. Darwin's description of the, 23
Barrier reefs of coral, 125
Basaltic rocks at Portland Bay, 121—157. At Cape Bridgewater, 157
Bathurst Island, 15

Bats, resort for, in a cave at Limestone Ridge, 366
Bay : —
— Grant, 153
— Guichen, 26, 52, 53, 150, 163, 169
— Lacepede, 159
— Portland, 26, 121, 157
— Rivoli, 159, 206
Beach Caves. See Caves
Beach terraces, formation of, 215. Levels of the, 218
Beasts of prey, Australian, 313
Bermuda, white mud of the reefs of, 93
Bight, the Australian, 116. Evidence of the nature of the, 118
Birds, scarcity of, in Australia, 140
'Biscay Country' or 'Dead Men's Graves,' 47
'Biscuits,' limestone, 42, 43. Their origin, 43
Blanche Caves, near Penola, 323. See Caves
Blandowski, M., his survey and maps of the three Lakes of Mount Gambier, 227
'Blow-holes' in the rocks at Guichen Bay, 170
Blue Lake, on Mount Gambier, 228. The four kinds of rock on the sides of, 229. Nature of the eruptions which have taken place in the crater of, 243, 250, 253. Its beautiful crystal water, 246. Its depth and flat bottom, 247. Subsidence coincident with volcanic disturbance, 251
Bombs, volcanic, 268
Bones, deposits of, on the banks of swamps, 54. And in crevices of the limestone rocks, 56. Theories respecting bones and caves, 302. How bones become preserved in rivers, 310. Bones found in caves always larger than those of species now in existence, 333. Bones in the first cave at Mosquito Plains, 334. Bones of Rodentia found, 335. Bones at the mouth of the Cave of Mount Burr, 357
Bonney Lake, 196. Description of it, 202
Boulders of trap rock in Grant Bay, 154
Brachiopoda, fossil, 79

INDEX. 393

Bridgewater Cape, 153. Rocks at, 154, 157
Bryozoa, fossil, of the limestone formation, 73. Meaning of the term, 73. Characters of Bryozoa, 73. Difference between it and the true coral, 73, 74. Table of fossil Bryozoa found by the author, 78. Age of Bryozoa, 88. Deposit of Mount Gambier derived from, 97
Bunce's description of the fossiliferous limestone of Tasmania, 122
Burr, Mount, 287. Character of the rocks, 287. Caves of Mount Burr, 353. Roots of trees hanging from the ceiling, 355. Mud at bottom, 356
Burra Burra copper mine, 5
Bursaria spinosa, of the ridges, 33

CAPE BRIDGEWATER, 153, 154, 157
— Grant, 153, 154, 163
— Jaffa, 162
— Jervis, 110
— Lannes, 152
— Otway, 121
— Paisley, New, 114, 116
— Yorke, 17
Caripe, the river, origin of, 317
Carpentaria, Gulf of, 15. Geological formation of, 17
Casuarina æquæfolia, the shea oak of the colonists, 30
Cats, native (Dasyurus Maugii), 56. Bones of, found in caves, 340.
Cave Station, origin of the immense basins of chasms at, 240
Caves, 299. Near Mount Gambier, 64. At Guichen Bay, 169. 'Blow-holes' of the, 170. Denudation and its effects, 299. Caves in the trap and limestone, 300. Four kinds of caves, 301. Points of resemblance between them all, 301, 302. Bones in caves, and theories respecting, 302. Caves made by fissures, 303. How bones come into them, 304. Parallel instances in South Australia, 305. Course of rivers in caves, 306. The Katavothra, or caves of the Morea, 306. The Swede's Flat, 309. How bones become preserved in rivers, 310. Why bones alone are found, 311. Caves which have been dens of animals of prey, 313. Sea-beach caves, 314. Paviland Cave, 314. Egress caves, 316. None in Australia, 316. The Cueva del Guacharo, 316. The mammoth caves of Kentucky and Tennessee, 319. Interest attaching to caves, 321. First cave at Mosquito Plains, 323. Second cave, 325. Third cave, 327. Dried corpse of a native, 329. Robertson's Parlour, 331. Connection between it and deeper caves, 332. Coralline fossiliferous limestone composing the rocks, 333. Bones found in caves always larger than those which are contemporaneous with man, 333. Bones in the first cave at the Mosquito Plains, 334. Bones of Rodentia found, 335. Other bones, 338. How the animals were embedded, 342. Result of inundation, 343, et seq. Cause of the signs of floods near the volcanoes, 347. How the limestone dissolved, 348. The theory of the bones found in caves being relics of the Deluge quite untenable, 349. Caves of Mount Burr, 353. Vansittart's Cave, 357. Mitchell's Cave, 359. The Drop-Drop Cavern, 360. Ellis's Cave, 363. Cave near Mount Shanck, 365. Caves at Limestone Ridge Station, 365. Bat Cave, 366, note. Other smaller caves, 366. Sir J. Mitchell's description of the caves in Wellington Valley quoted, 373
Cellepora gambierensis, 74, 85, 97. Where mostly found, 91
Cephalopoda, fossil, found by the author, 80
Cerithium in the recent limestone, 190
Cetacea, fossil remains of, found on the banks of the Murray, 80
Chalk formation, origin of the, 13, 14. Similarity of the mud of coralline reefs to chalk, 92, 93. Origin of the chalk of England and France,

96. Extent of the chalk formation of Europe, 100
Chara, remarkable growth of the, in a fresh-water lagoon, 53
Cibotium Billardieri, 359
Cidaris, fossil spines of, 81
Cirripedia, fossil, 78
Clarence Strait, 15.
Clune's gold mine, near Ballarat, 297
Clypeaster, 77
Coal, beds of, in Victoria, 22. In Western Australia, 22
Const, the geological character of a country told from the appearance of the, 13
Colac, Lake, deposit of bones on the banks of, 55
Conchifera, fossil, found on the banks of the Murray, 105
Conclusions, summary of, 368
Concretions of lime and sand of the Cape Jaffa reefs, 162, 163. Their origin, 166. Not fossil trees, 167
Conglomerate, tenacious, found in Grant Bay, 159
Coorong, the, 195, 364. Description of, 197
Copper mines of Burra Burra, 5. And of Kapunda, 6
Coral, moss. *See* Bryozoa
Corals, found on Bald Head, 115, 116. Depth of sea in which they can live, 89. How deposited, 90. Corals of Mosquito Plains, 133
Coral islands, or atolls, 13, 14. Mr. Darwin's theory of the formation of the, 125
Coralline limestone, 86
— reefs, 91. The mud derived from, 92. Description of a reef, 94. Difficulty as to the nature of the coral, 99. Extent of the formation, 99. Class of reefs to which the coralline crag belongs, 127. Cessation of the coralline formation, 148
Corethrostylis Schultzenii, 53
Cordillera, Australian, 18
Cornwall, the sand formation of, 184
Correa cardinalis of the heath country, 37
Crag, lower, of England, probable identity of the, with the tertiary deposits of South Australia, 85, 86. Features of the crag, 86
Crag, upper, description of the, 150. Its extent, 150. Derived from an ocean current, 151. Material of which it is composed, 152. Upper crag at Cape Lannes, 152. Of Grant Bay, 155. Other localities in Australia in which it is found, 159. Its variable capability for resisting the action of the atmosphere and sea-water, 161. The concretions called fossil-trees, 163. History of the deposit of the upper crag, 173. Evidence of its partial destruction, 174. Want of uniformity in its thickness, 174. Causes of this, 175. Raised, 175. Its subsequent removal, 176. Its age, 176, 177. Compared with the coralline crag of Suffolk, 177. Absence of fossils in the South Australian crag, 178
Craters. *See* Volcanoes. Of subsidence, 284
Cray-fish of the plains, 43, 51. Holes made by the, 43
'Cribbage-pegs, fossil,' 81
Cristellaria of the limestone rock, 71
Cruziana cncurbita, found at Nuriootpa, 21, *note*
Cyclops vulgaris in the swamps, 52
Cypris, the, in the lakes of South Australia, 52

DARWIN, Mr., his theory of the limestone formation, 125
Dasyurus Maugii, or native cat, bones of the, found in caves, 340. Found in Wellington Valley, 382
'Dead Men's Graves,' or 'Biscay Country,' 47. Origin of the graves, 47
Deluge, the bones found in caves no evidence of the, 349
Denudation and its effects, 299
Desimadæ in the swamps, 52
Diamonds found in the South Australian range of hills, 21
Diatomaceæ in the swamps, 52. Sand composed of the frustules of, 53
Dillwynia floribunda, 37
Dimyaria, fossil, found by the author, 79

INDEX. 395

Dingo, the Australian, 313
Diprotodon, bones of, found in Wellington Valley, 381
Dismal Swamp, the, of the South-Eastern District, 27
Disturbance, common in the northern, but uncommon in the southern hemisphere, 225, 226
Dolerite found in the craters of Mount Gambier, 258
Dolomite of the 'Biscay Country,' 48. Origin of, 48
Drainage, subterranean, 364
Drop-Drop Cavern, 360
Droughts, periodical, of Australia, 145
Dunes, sand. *See* Sand Dunes

EARTH, analogy of the present state of its crust with former geological epochs, 136
Earthquakes, and the upheaving of the South Australian coast, 213
Echinidæ, beds of fossil spines of, 81
Echinoidea, 77
Echinolampus, 77
Edward, Lake, 283. Evidences of volcanic action, 283
Edward, Mount, 288
Egress caves. *See* Caves
Eliza, Lake, formation of salt taking place in, 69, 195. Description of, 199
Ellis's Cave, 363. An underground channel of drainage, 364
Entomostraca, 78
Entromostraca brachiopoda, 73
Epacris impressa of the heath country, 37
Eremacausis, instance of, in the caves of the Mosquito Plains, 329. Causes of, 330
Etna, bottom of the crater of, 272
Eucalyptus dumosa, or mallee scrub, 33
Eucalyptus fabrorum, or stringy bark, 31
Eucalyptus resinifera of the ridges, 30
Euro, bones of the, found, 361

FAIRY, PORT, 121
Fascicularia, found on the South Australian coast, 187, 188, *note*

Fern, the common Australian, 30
Fern-tree growing in a cave from bottom to top, 358
Flinders, Captain, his description of the coast of South Australia, 114
Flints, layers of, in the limestone formation, 64. Origin of, 65. Separation of silica, 65. Origin of the beach terraces, 217
Flora of South Australia, 36, 37. Its correspondence to the secondary period, 139. Beauty of the flora of Australia, 267
Fish in a lake highly impregnated with salt, 54 Fossil remains of, found by the author, 80
Fissure caves. *See* Caves
Foraminifera in limestone formation, 70
Fossil cliffs of the great Australian Bight described by Mr. Eyre, 386
Fossiliferous rock of part of the coast of Australia, 18. Extent of, in South America, 225
Fossils not found in the South Australian mountains, except at one or two points, 20. Fossils found in Victoria observed to agree with those of Europe, 22. Fossils of the upper limestone, 61. Singular fossils found near Penola, 75. List of fossils found on the banks of the Murray, by Captain Sturt, 105. Fossils to the eastward of the boundary line between Victoria and South Australia, 120. Evidence of transport of fossils, 134. Absence of fossils in the Guichen Bay deposits, 152, 153, 178
Fossils, list of, found in South Australia: —
— Asteroidea, 78
— Astræa, 187
— Astro-Pecten, 83
— Brachiopoda, 79
— Bryozoa, 73, 78
— Cellepora gambierensis, 74, 85, 91, 97, 105, 333
— Cephalopoda, 80
— Cerithium, 190
— Cidaris, 81
— Cirripedia, 78
— Clypeaster, 77

396 INDEX.

Fossils — *continued* : —
— Conchifera, 105
— Cristellaria, 71
— Cruziana cucurbita, 21, *note*
— Dimyaria, 79
— Echinidæ, 81
— Echinoidea, 77
— Echinolampus, 77
— Entomostraca, 78
— Entomostraca brachiopoda, 73.
— Fascicularia, 187, 188, *note*
— Foraminifera, 70, 77, 189
— Gasteropoda, 79
— Globigerina bulloïdes, 72, *note*
— Glossopteris Browniana, 22
— Lunulites, 75
— Mollusca, 106
— Monomyaria, 79
— Monostega, 71
— Murex asper, 84
— Nautilus ziczac, 83
— Operculina arabica, 72, *note*
— Pectens, 74, 76, 333
— Pecten Jacobæus, 160
— Pentamenus oblongus, 20
— Pisces, 80
— Polyozoa, 84
— Radiata, 105
— Salicornaria, 105
— Spatangus Forbesii, 75, 83, 165
— Terebratula compta, 74, 121
— Trochus, 77
— Turritella terebralis, 83
— Venus exalbata, 190
— Zoophytes, 77
Francis, St., Isles of, formation of the, 119
Fringing reefs of coral, 125

GAMBIER, MOUNT, native wells near, 63. Caves near, 64. Fossils found at, 72 and *note*, 74. Upper crag formation, 160. Oyster-shell bed at, 160, 161. Description of the extinct crater of, 227. Captain Sturt's observations, 227. M. Blandowski's maps of the three lakes, 227. Description of the three lakes, 228. And of the crater walls, 235. The oldest crater obliterated, 237. What kind of eruption has taken place to produce the appearances presented, 237. Its small extent, 239. Causes of the non-appearance of ejectamenta on the east side of the crater, 240. The promontories or ridges jutting out from the walls, 241. Account of an active volcano resembling Mount Gambier, 248. Underground flow of the lava of Mount Gambier, 250. Number of craters at, 253. Three periods of their activity, 254. Age of the crater, 255. Fossils in the rocks of, 255. Peculiarities of the strata of, 255. Period of the last eruption, 256. Minerals found in the craters, 258. Review of the past of Mount Gambier, 259. Resumé of its geological features, 263. Beautiful view from its summit, 264. Connection between Mounts Gambier and Shanck, 279. Hilly country round Mount Gambier, and probable causes, 280

Gasteropoda, fossil found by the author, 79

Geology, importance of the science of, to man, 2—4

George, Lake, 195. Description of it, 201

German Flat, the quagmire so called, 203. As seen from Mount Muirhead, 204.

Glaciers, traces of the action of, in the South Australian chain of mountains, 20

Glenelg river, 15, 27, 28, *note*. Character of the plains through which it runs, 41. Upper crag formation at the mouth of the, 159. Evidence it affords of the upheaval of the coast, 209

Globigerina bulloïdes, 72, *note*

Glossopteris Browniana, in the coal beds of Victoria, 22

Goa, Mount, in the Sandwich Islands, description of the volcano of, 248

Gold found in the South Australian range, 21. Trap rock not always an indication of the existence of, 297. History of the formation in which gold is found in Victoria, 297. Clunes

INDEX. 397

Mine, near Ballarat, 297. Gold of South Australia, 298
Graham, Mount, black mud swamp at the foot of, 210, 287, 364
Granite rocks in the Murray river, 118, 119. In South Australia, 298
Grant Bay, extent and boundaries of, 153, 154. Boulders of trap rock in, 154. Strata of the coast described, 154, *et seq.*
Grant, Cape, 153. Rocks at, 154. Twisted concretions in the cliffs at, 163. Their origin, 164
Guacharo, the Cueva del, Humboldt's description of, 316
Guichen Bay, 26. Fresh-water lagoon near, 53. Thick growth of the common Chara in the, 53. Description of the rocks at, 150. Their extent, 150. Their origin, 151. Material of which the rocks are composed, 152. Wild scenery of the coast, 153. Absence of fossils at, 152, 153. Twisted concretions in the cliffs at, 163. Their origin, 164. Caves at, 169. 'Blowholes' at, 169. Sand-hills at, 171. Genera of shells found at, 194
Gum of the Eucalyptus resinifera, 30. Of the wattle, 30
Gum-trees of the ridges in South Australia, 29
Gypsum, crystals of, in the mud of the lakes, 69

HALF-WAY GULLY, probably the remains of a reef, 131
Hawdon, Lake, 193. Description of, 195, 198
Heat, volcanic sand acting as a nonconductor of, 245. Nasmyth's experiment, 245
Heath, character and extent of, in South Australia, 32
Henderson Island, Lyell's description of, quoted, 156
Hills, character of the six chains of, 214
Honeysuckle country, the, of South Australia, 33, 42
Hopeless, Mount, 110
Hyæna caves, 313

Hypsiprymnus, bones of, found in Wellington Valley, 380

ICARI, existing on living polypifers, 93. Igneous rocks of South Australia, 226, *et seq.* *See* Volcanoes
Infusoriæ, the, in the swamps, 52. Silica in the shells of, 65.
Inundations, former, in South Australia, 343, 345
Investigator's Straits, 112
Iron pyrites in the upper limestone, 67
Island, Bathurst, 15
— Henderson, 156
— Kangaroo, 15, 112
— Keeling, 93
— Julia Percy, 292
— Melville, 15
Islands, coral, 13, 14
— St. Francis, 119

JAFFA, CAPE, reef of rocks at, 162. Upheaval of the reefs at, 207
James, Mr. G. P R., his romantic account of Mount Gambier, 227.
Jervis, Cape, 15, 16, 110. Geological formation of, 17
Johnstone's Station, cave at, 366
Julia Percy Island, an extinct volcano, 292

KANGAROO ISLAND, 15, 112
Kangaroo bones in the swamps, 55. Those found in the Wellington Valley much larger than those of any existing species, 341. Bones of a large kangaroo, 361
Kapunda, town of, 6. Mine of, 6
Katavothra of the Morea, their similarity to the caves at Mosquito Plains, 344
Keeling Island, white mud of, 93
Kentucky, mammoth caves of, 317
Kilauea, description of the active volcano of, 248
Kirkdale Cave, in Yorkshire, 312
Kooringa, rise of the town of, 5
Kosciusko, Mount, its height, 19, *note*

398   INDEX.

LACEPEDE BAY, upper crag formation at, 159
Lagoons, fresh-water, of the coast, 196. Limestone, formation of the, 197
Lake Albert, 204, 205
— Alexandrina, 204
— Blue, 228, 243, 250, 253
— Bonney, 196
— Colac, 55
— Coorong, the, 195, 197
— Edward, 283
— Eliza, 69, 195
— George, 195, 201
— Hawdon, 193, 195, 198
— Leake, 283
— Middle, 230
— near Mount Shanck, 266
— Roy, 51
— St. Clair, 195, 199
— Torrens, 110, 113, 117
— Valley, 231, 240
Lakes, two remarkable for their deposits, 53. Crystals of gypsum and natron found on the shores of the, 69. Lakes on the coast of South Australia, remarks on the, 195, et seq.
Lannes, Cape, upper crag at, 152
Lap-lap, the fish of the swamps so called, 50
Lava of Mount Gambier, 250. Of Mount Shanck, 267
Lawrence Rock, an extinct crater, 292. Nature of the strata of, 292. Overlaid by crag, 293.
Leake, Lake, 283. Character of the banks, 283. Evidences of volcanic action, 283
Leake's Bluff, 264, 284. Geological character of, 285. Fault at, 285, 286
Limestone ridges, 30
Limestone Ridge Station, caves at, 365
Limestone 'biscuits,' 42, 43. Their origin, 44
Limestone, the upper, 60. Horizontality of the beds, 60. Distribution of fossils in the, 61. Table of fossils found in the, 77, et seq. Age of the beds, 82. Position of the beds, as reported by Dr. Busk, 84. Their probable identity with the lower crag of England, 85, 86. Similarity of limestone to coral rock, 97. Extent of the formation in South Australia, 100, et seq. Boundaries of the district, 103. Fossiliferous limestone on each side of Lake Torrens, 113. Extent of the formation to the eastward, 120. The Tasmanian beds, 122. Origin of the limestone formation, 124. Subsidence, 124. Darwin's theory, 125. Probable remains of reefs at Swede's Flat and Half-way Gulley, 131. Bed of oyster-shells on the top of nearly every limestone cliff, 160, 161. Limestone concretions at Cape Jaffa Reef, 162. Features of the upper limestone, 189. Shell deposits of this formation, 190. The limestone of Guichen Bay, 194. Of the lagoons, 197. The six chains of hills, 220. Causes of caves in limestone, 300. See Caves
Limestone, coralline, 86. Coralline fossiliferous limestone composing the rocks of the caves, 333
Limnea stagnalis in the swamps, 51
Lincoln, Port, metamorphic rocks about, 119
Lithodomi, 214. Borings of, at Mount Gambier, 160. And at other places, 174
Lofty, Mount, 20. Rocks of, 59
Lunulites found near Penola, 75

MACROPUS TITAN, bones of the, 357, note
Macropus Athos and Titan, bones of, found in Wellington Valley, 378
Magnesia, large quantity of, in the 'Biscay Country,' 48. Magnesian fermentation, 48
Magpie, native (Gymnorrhina leuconota), 56
Mallee scrub of South Australia, 33. Its character and extent, 33
Mammalia, low organisation of the, of Australia, 140
Mammoth caves of Kentucky and Tennessee, 317
Man, developement of an approach to a more complex organisation ending in, 137

Marsupialia of Australia, 140
M'Arthur, Mr. Donald, singular underground sounds heard at his station, 57
M'Donnell, Port, sheets of flint layers at, 64
M'Intyre, Mount, 264, 288
Melaleuca, the tea-tree of the colonists, 228
Melbourne, its appearance in 1850 compared with that of the present day, 3. Earthquakes at, 213
Melicerita, genus of, peculiar to the tertiary beds of South Australia, 85
Melville Island, 15
Meredith's Station, cave at, 366
Mesembryanthemum of the coast at Guichen Bay, 153
Metals found in the South Australian range, 21
Middle Lake, of Mount Gambier, 230. Nature of the eruption from the crater of, 242
Minerals found in the South Australian range, 21. Found in the craters of Mount Gambier, 258
Mitchell's Cave, 359. Its deep sea-blue pool of clear water, 359. Sand-pipes, 359
Moleside rivulet, in Tasmania, 122. Character of the country near the, 122
Mollusca, fresh-water, of the swamps of the South-Eastern District, 51
Mollusca, fossil, found on the banks of the Murray, 106
Monomyaria, fossil, 79
Monostega of the limestone rock, 71
Morea, caves in the, 306
Mosquito Plains, absence of trees in parts of the, 49. Probable causes of this, 50. Strata of the, 76. Coral of the caves, 91. Fossils of the, 133. Caves at, 323. First cave, 323. Second cave, 325. Third cave, 327. Dried corpse of a native there, 329. Robertson's parlour, 331. Connection between it and deeper caves, 332. Bones in the first cave, 334.
Mosquito Swamp, 364
Moss coral. *See* Bryozoa
Mount Burr. 253, 287

Mount Edward, 288
— Gambier, 63, 64, 72, 227, 290
— Graham, 210, 287, 364
— Hopeless, 110
— Lofty, 20, 59
— M'Intyre, 288
— Muirhead, 204, 286
— Napier, 157
— Rouse, 157
— Shanck, 63, 250, 261, 290, 365
Mountains, the Australian Cordillera, 18. The South Australian Chain, 19. Their dimensions, 20
— Barossa, 110
— chain from Cape Jervis to Mount Hopeless, 110
Muirhead, Mount, 286. Evidences of volcanic action, 287
Munbannar, caves near the town of, 365
Murchison, Sir Roderick, his prediction as to the auriferous regions of Australia, 4
Murex asper, 84
Murray river, 27. Fossil bones of Cetacea found on the banks of the, 80. Capt. Sturt's survey of the, 104. List of fossils found by Capt. Sturt on the, 105. Description of the cliffs, 106. Country to the north of the river, 109. Extent of the formation in a westerly direction, 110. Fossils found on the, 133. Lakes lying at the mouth of the river, 204. Evidence it affords of the upheaval of the Australian coast, 209. Granite of the bed of the, 298
Murray Scrub, the country so called, 111
Myrmecobius, the, 339

NAMES of places, native, 365, *note*
Napier, Mount, 157
Natron found on the shores of the lakes, 69
Nautilus ziczac, fossil, of England and South Australia, 83
Negro, Rio, great sandstone plateau of, 222

Norfolk Island, flora of, 139
Nummulites, absence of, in the upper limestone, 75

OAK, the shea, of South Australia, 30
Olivine found in the craters of Mount Gambier, 258
Operculina Arabica, 72, *note*
Otway, Cape, 121
Oxyrrhinus Woodsii, of Mount Gambier, 133
Oyster-shells, abundance of, on the tops of the cliffs of the Murray river, 109. Bed of, on the top of nearly every limestone cliff, 160. Period of the deposit, 161.

PACIFIC OCEAN, coral islands of the, 125. Subsidence of a large portion of the bed of the, 135
Paisley, New Cape, 114, 116
Paludina, banks of the shells of, at Lake Roy, 51
Pampean formation, extent of the, 225
Paviland Cave, 314
Peak of Derbyshire, 322
Pectens, 74, 76
Penola, limestone ridge at, 31. Freshwater mollusca found in the neighbourhood of, 51. Singular fossils found near, 75. Caves near, 323. *See* Caves
Pentamenus oblongus, found near Adelaide, 20
Phalangesta, bones of, found in Wellington Valley, 380
Phascogale penicillata, or native squirrel, bones of, found in caves, 339
Phascogale pygmæa, 339
Phascolomys, bones of, found in Wellington Valley, 381
Plain, extensive, of the South-Eastern District, 27
Plain, strata of the South-Eastern District, 58

Pleiocene strata of Rome, Sir C. Lyell on the, 87
Polyozoa of the tertiary beds, 84
Porphyry hills in the north of the South Australian District, 27
Portland, town of, its cold and sombre appearance, 26
Portland Bay, 26. Coralline limestone at, 121. Fossils at, identical with those of Mount Gambier, 121, 133. Basaltic rocks on the tops of the cliffs at, 157
Portland, Victoria, veins of soapstone at, 65. Shell deposits near, 191, Evidence of subterranean volcanic action at, 291
Post-Pleiocene formation. *See* Limestone, upper.
Primary formation of the eastern and western sides of Australia, 16, 18
Pteris esculenta, or common Australian fern, 30, 358
Punch-bowl basin of Mount Gambier, 232. Conjectures as to its origin, 233
Pyrites, iron, in the upper limestone, 67

QUAGMIRE of German Flat, 203

RADIATA, fossil, found on the banks of the Murray, 105
Reedy Creek, an embryo river, 210—212
Reef, the great Barrier, of the west side of Australia, 22. Darwin's description of the, 23
Reefs, Cape Jaffa, 162. Singular appearance of some of them, 162
Reefs, coralline, 91. Description of a reef, 94. Of the Pacific, different kinds of, 125. Why no remains of reefs are found, 130. Probably some remains at Swede's Flat and Half-way Gulley, 131. Causes fatal to the progress of, 148. Subsequent history of, 181. Upheaval of the reefs at Rivoli Bay, 206. And at Cape Jaffa, 207

Ridges formed on the east shores of the swamps, 29. Different kinds of, 30. Vegetation of, 30.

River:—
— Glenelg, 15, 27, 28, *note*, 41, 159, 209
— Murray, 27, 80, 104, 209, 298
— Wannon, 210
— Wimmera, 27

Rivers of South Australia, proof they afford of the upheaval of the coast now going on, 208. A river in an early stage of developement, 210. Underground rivers, 122, 306. The present banks of rivers formerly the beds, 311

Rivoli Bay, upper crag formation at, 159. Recent alterations in the soundings at, 206

Robe Town, its cheerful aspect, 26

Rocks of the South-Eastern District, 58. Character of the rocks of South Australia, 58.

Rodents of Australia, 140

Rome, Sir C. Lyell, on the Pleiocene strata of, 87

Rouse, Mount, 157

Roy, Lake, banks of, fresh-water mollusca at, 51

SALT CREEK, 364
Salt-pans, or 'Salinas,' in the South-Eastern District, 69. Origin of these deposits, 69. Lake Eliza, 69

Salt, rock, found in the upper limestone, 67. Suggestions as to its origin, 63

Sand of the South-Eastern District, 37. Different kinds of, 37. Theory of the origin of sand, 39. Composed entirely of the frustules of Diatomaceæ, 52

Sand dunes, 167. Immense number of, round the coast of South Australia, 166. Contain no perfect fossils nor signs of stratification, 167. Sand-hills at Guichen Bay and Cape Bridgewater, 171. Interminable sand dunes of the coast of Australia, 183. Character of the sand, 183. Its extent inland, 183. The sand formation of Cornwall, 184. Origin of the sand on the South Australian coast, 187. And of calcareous sandstone, 188. Its composition, 188. Shell deposits, 190. Extent of the sandy coast, 219. Different kinds of the sand, 219. Immense size of some of the dunes, 219. Appearance of the dunes on windy days, 219. Encroachment of the sand on the land, 220. No indication of the formation of the sand into stone, 220. Trees buried in the sand, 221. Sandstone formation of the coast below the sand, 222. Why so associated 222

Sandhurst, rise of, 4

Sand-pipes of the upper limestone, 61. Origin of, 62. Those exposed in Mitchell's Cave, 359

Sandstone, ferruginous, of Cape Yorke and Arnhem's Land, 17. Calcareous, of Guichen Bay, 150. Of Grant Bay, 155. Formation of sandstone on the coast below the sand, 222. Why so associated, 223

Sandy ridges, 31. Vegetation of, 31

Scenery, dependence of, on geology, 25

Scrub, character and extent of, in South Australia, 33

Sea, temperature of the, in former times greater than now, 134

Sea-water, colour of, 179

Secondary rocks, absence of, in Australia, 140. State of the earth after the secondary period in Europe, 141

Shanck, Mount, native wells near, 63. An extinct volcano, underground flow of the lava of, 250. Description of the extinct volcano of, 261, 263, *et seq.* Country around it, 263. Beautiful little lake near, 266, 269. Volcanic bombs, 268. The great cone of Mount Shanck, 269. Its steep ascent, 269. View of the crater from the top, 269. Its shape, 270. Remains of a former crater on the east side, 270. Trees growing inside and outside the basin of the older crater, 271. Composition of the more recent crater wall, 272. Bottom of the crater, 272, 273. Peculiarity in the layers of ash, 273. No apparent outlet for the lava, 273. Stream of lava on the north, 273. Curious mode in which it is heaped, 274. A current from the ancient crater, 275. Causes of the heaping up, 276. Connection between Mount Gambier and Mount Shanck, 279. Dismal view of the coast line from Mount Shanck, 281. Cave at, 365

Shea Oak, the, 30

Shell deposits on the coast, 190. Localities in which they occur, 191, et seq. Their extent inland, 193. Observations on those found at Guichen Bay, 194

Silica. See Flint

Silurian rocks from Cape Yorke to Port Phillip, 18

Soapstone, veins of, at Portland, in Victoria, 65

Soil, different kind of, on the South Australian ridges, 30. Varieties of, in the South-Eastern District, 38

Sounds, singular, of the swamps which have an underground drainage, 57

Spatangus Forbesii, 75, 83

Spencer's Gulf, 15, 16, 112. Evidence at, of periods of upheaval and of rest, 213

Sponges, silica in, 65

Squirrel, native, bones of, found in caves, 339

Stalactite, an enormous, 324. Of the second cavern at Mosquito Plains, 326

Stalactites, how formed, 348

Stalagmites, 323

St. Clair, Lake, 195. Description of, 199

Steam craters of volcanoes, 269

Stone Hut range of hills, shell deposits on, 193

St. Paul, Island of, causes of the difference in the height on the eastern and western sides, 241

Stratification, singular, of the coast at Guichen Bay, 153. At Grant Bay, 154. No signs of stratification in the sand dunes, 167

Sturt, Captain, his survey of the river Murray, 104. His account of the formation to the north, 112. His observations on the extinct crater of Mount Gambier, 227

Swamps of the South-Eastern District, 27. The Dismal Swamp, 27. Their localities and peculiarities, 28. Fertile ridges, 29. Swamps in the grassy plains, 41. Peculiarities of swamps, 50. The fish Lap-lap, 50. The cray-fish of the plains, 43, 51. Fresh-water mollusca, 51. Two swamps remarkable for their deposits, 53. Bones on the banks, 54. Singular sounds connected with the swamps which have an underground drainage, 57. The German Flat, 203. Black mud swamp at the foot of Mount Graham, 210. Mosquito Swamp, 364

Swede's Flat, probably some remains of reefs at, 131. Description of the, 309

Subsidence of the limestone formation, 124, et seq. Subsidence coincident with volcanic disturbance, 251. A crater subsidence theory, 252

Superior, Lake, calcareous sandstone near, 222

TAPLEY'S HILL, near Adelaide, formation of, 208

Tatiara, probable remains of reefs in the, 131

Tasmania, mineral wealth of, 6. Fossiliferous limestone formation in, 122

Tea-tree (Melaleuca paludosa), 36, 222

Tennessee, mammoth caves of, 317

Terebratula compta, 74

Terraces, sea-beach, formation of, 215. Levels of the, 218

Tertiary beds in the centre of Australia, 16, 18. Tertiary formation and its fossils, 60, et seq. Tertiary strata of Rome, 87. Of Asia Minor, 88.

Tetratheca ciliata of the heath country, 37

Torrens, Lake, 110, 113, 117

Trap rock of part of the coast of Australia, 18. Boulders of, in Grant

Bay, 154. Trap rocks of the coast of the Bay, 156. Amygdaloidal character of the trap, 158. Trap rock of Leake's Bluff, 285. Trap rock not always an indication of the existence of gold, 297. Causes of caves in trap rock, 300

Trees, distribution of, in the South-Eastern district, 49. Fossil, of the crag, not trees, but magnesian limestone infiltrations, 165—168. Trees buried in the sand, 221. Not fossilised, 222

Trochus, 77

Tunicata, fossil, found on the banks of the Murray, 105

Turritella terebralis, 83

UPHEAVAL of the Australian coast, 203. Singular instances within the last few years, 207. Six periods of rest in the upheaval, 215

VALLEY LAKE of Mount Gambier, 231. The crater walls surrounding the lake, 231. Mode of eruption of the crater, 241. Subsidence coincident with volcanic disturbance at, 251

Van Diemen's Gulf, 15

Vansittart's Cave, 357. Wonderful clearness of the water in, 357. No bones perceptible at, 358. Fern-trees at the entrance, 358

Vegetation of the South Australian ridges, 30

Venus exalbata in the recent limestone, 190

Vesuvius, bottom of the crater of, 272

Victoria, gold fields of, 4. Geological government survey of, 7. Geological results of the gold-digging in, 21. Its fossils agreeing with those of Europe, 22. Coal beds of, 22. Boundary line between, and South Australia, 120. Immense volcanic district of, 290. History of the formation in which gold is found, 297

Vincent, Gulf St., 15, 111

Volcanoes, no active and few extinct, in Australia, 224, 222. Description of the extinct crater of Mount Gambier, 227, et seq. A theory of crater subsidence, 252. Description of Mount Shanck, 261, et seq. Dissimilarity in volcanoes, 261. Reasons for multiplying the records of volcanic action, 262. Volcanic flora of Australia, 267. Volcanic bombs, 268. Craters of steam and ashes by the sides of volcanoes, 269. Appearances presented by the bottoms of the craters of extinct and active volcanoes, 272. Curious mode in which the lava stream of Mount Shanck is heaped, 274. The smaller volcanoes, 282. Crater lakes, 283, et seq. Connection which existed between the volcanoes of the South-Eastern District, 288. Supposed causes of volcanoes, 289. Igneous reservoirs, 290. Line of disturbance probably connected with the Victoria volcanic district, 290. Theory of the existence of volcanoes in a recent tertiary period, 291. Evidence of subterranean volcanic action of Portland, 291. Lawrence rock, 292. Julia Percy Island, 292. Evidence of periods of disturbance and rest, 293. Von Buch's theory of craters of elevation, and the controversy which resulted, 294

Volcanoes, extinct :—
— Gambier, 63, 64, 72, 227, 290
— Napier, 157
— Rouse, 157
— Shanck, 63, 250, 261, 365
— smaller craters, 282

WALES, New South, gold-fields of, 5. Geological examination of, 7

Wannon river, evidence it affords of the upheaval of the coast, 210

Water-level, facts with reference to the, in South Australia, 179.

Wellington Valley, Sir Thomas Mitchell's description of the caves at, quoted, 373. Professor Owen's report on the bones found in the caves, 378.
Wells, native, 63. Origin of the, 63, 64.
Well-shaped holes in the country between Mount Gambier and Mount Shanck, 265
Wimmera river, 27
Wombat bones in the swamps, 55

XANTHORRHŒA AUSTRALIS, 36, 140

YORKE'S PENINSULA, 15, 111. Geological formation of, 17
Yorke, Cape, geological formation of, 17

ZAMIÆ, 139
Zoophytes, fossil, 77

www.ingramcontent.com/pod-product-compliance
Lightning Source LLC
Chambersburg PA
CBHW020542300426
44111CB00008B/765